OPPORTUNITIES IN DAM PLANNING AND MANAGEMENT

OPPORTUNITIES IN DAM PLANNING AND MANAGEMENT

A COMMUNICATION PRACTITIONER'S HANDBOOK FOR LARGE WATER INFRASTRUCTURE

LEONARDO MAZZEI

LAWRENCE J. M. HAAS

DONAL T. O'LEARY

THE WORLD BANK
Washington, D.C.

© 2011 The International Bank for Reconstruction and Development / The World Bank
1818 H Street NW
Washington DC 20433
Telephone: 202-473-1000
Internet: www.worldbank.org

1 2 3 4 :: 14 13 12 11

ISBN: 978-0-8213-8216-5
eISBN: 978-0-8213-8279-0
DOI: 10.1596/978-0-8213-8216-5

Cover photo: Leonardo Mazzei

Library of Congress Cataloging-in-Publication Data
Mazzei, Leonardo, 1973-
Opportunities in dam planning and management: an infrastructure practitioner's handbook / prepared by Leonardo Mazzei with Lawrence Haas and Donal T. O'Leary.
        p. cm.
    Includes bibliographical references and index.
    ISBN 978-0-8213-8216-5 (alk. paper)—ISBN 978-0-8213-8279-0 (electronic : alk. paper)
    1. Dams—Design and construction. 2. Infrastructure (Economics)—Decision making. 3. Project management. I. Haas, Lawrence J. M. II. O'Leary, Donal. III. Title.
TC540.M463 2010
363.6'1—dc22
                                    2010032690

Contents

Tables

Preface

The past few decades have seen major advancements in the delivery of environmentally responsible and socially acceptable infrastructure in different governance settings. However, the yardsticks needed to measure progress and define good practice continue to be redefined.

The emergence and wider acceptance of dialogue and partnership approaches that incorporate multiple stakeholders are trends that characterize beneficial changes in dam planning and management. Several motivations are driving these trends.

The first is to involve interested and affected people as partners in development decisions, thereby realizing the many benefits that inclusive approaches are acknowledged to bring. A second motivation is to inform and legitimize what are ultimately political decisions on proceeding with major dam developments, balancing the economic, social, and environmental performance aspects with the implementation and operations phases of dams. Project experience recognizes that this balance evolves over time, is contextually defined, and can be adjusted through stakeholder processes.

This dimension of development practice requires integrating communication strategy within the entire project process. Mechanisms for dialogue are needed to bring government, civil society, and private sector actors together to advocate and deliver the various governance reforms and anticorruption measures that influence the overall development performance of dams, especially to ensure that the benefits and costs of dam developments are equitably distributed in societies—and perceived to be so.

Several multifaceted concerns also drive the need to embed modern communication thinking in dam planning and management more generally. At the center is the concept that large dam developments should be approached holistically, as wider development interventions, rather than narrowly, as physical assets delivering water and energy services. The local communities that host dam projects and the river users who rely on natural resource systems hold this view strongly. Their attitudes are revealed powerfully when interested and affected parties engage in two-way communications at each stage of the infrastructure project cycle.

For a communication practitioner, developing communication practices in dam planning and management means informing the project design, especially the social and environmental dimensions of sustainable performance as it is defined and perceived by stakeholders. It also means having a clear picture of the political and cultural dynamics and other behaviors that hinder or advance the formation of vital partnerships and of how to nurture them and to ensure power relations are balanced. Implementation of this full interpretation of the development of communication practices requires analytical tools to understand the values, fears, and expectations that lay beneath the varied, and too often highly polarized, perceptions of risk, which, when left unattended, form the fault lines of conflict.

On a practical level, the development of communication practices means adequately investing in communication at all project stages aimed at building functional stakeholder partnerships, working with the media, and ultimately spending less on crisis management. It also means understanding the political economy of dam decisions, as well as the steps needed to create trust, good faith, and the space within which sustainable, pro-poor, pro-environment solutions can be incorporated. Equally, it means advocating the relevant good development practices promoted by the World Bank and undertaking advocacy outreach to all stakeholders, from government policy makers to families affected by dam projects.

This handbook responds to the call by the World Bank's Development Committee for infrastructure and communication practitioners to collaborate on more comprehensive communication strategies for all stages of the Bank project cycle widely endorsed by international stakeholders. It is timely for several reasons:

- Large dams and hydropower are climbing back to the top of the global development agenda because of the growing appreciation of the manifest linkages among water, environment, energy security, and climate change. In response, the World Bank is scaling up its own portfolio of dam projects, including initiating new projects and rehabilitating and reopening the operations of existing dams.
- The operational or development communication practice offers a range of new tools and techniques suited to today's challenges and complexity in dam planning and management.
- The democratization of development processes increases the pressure for inclusive decision-making processes and the full translation of policy reforms into dam projects.
- The growing recognition of corruption is acting as a safeguard against the manifest effects of corruption on the provision of infrastructure. New anticorruption tools are available but underutilized. At the same time, communication is needed to underpin a shift to more comprehensive approaches to risk management.

Structures that promote more dialogue respond to these new challenges and equip project task teams with the tools they need to better manage stakeholder expectations and to explore solutions through partnerships.

This handbook is one of three products prepared under an initiative supported by the Bank–Netherlands Water Partnership Program (BNWPP). This volume, *Opportunities in Dam Planning and Management: A Communications Handbook for Large Water Infrastructure*, offers task managers and infrastructure practitioners the basic information they need to benchmark value-added improvements in communication within the World Bank's own portfolio of dam projects. The other two products are two World Bank working papers: *Berg Water Project: Communications Practices for Governance and Sustainability Improvement* and *Lesotho Highlands Water Project: Communications Practices for Governance and Sustainability Improvement.*[1] These papers provide an in-depth assessment of two dam projects in Africa, looking at governance, sustainability, and communication to draw lessons. Synopses of the two case studies appear in appendixes A and B of this handbook.

This handbook is aimed primarily at two groups of practitioners: (1) those already conversant with governance and anticorruption concepts and communications support in World Bank operations, and (2) developing country planning process or dam project managers—that is, practitioners less familiar with the language or concepts associated with these topics who work on World Bank–supported dam projects or on dam planning and management concerns in different country settings.

For communication specialists, the handbook contributes to the dialogue on ways to modernize communication in dam planning and management. Some of the concepts and illustrations of good practice cited in the handbook are also relevant to other aspects of infrastructure, such as communication support to advance governance and anticorruption (GAC) efforts along the infrastructure project cycle.

Objectives of the Handbook

This handbook seeks to advance efforts to evolve and benchmark value-added improvements in communication practices in dam planning and management. Another objective is to improve awareness of the opportunities to exploit synergy in linking improvements in governance, sustainability, and communication. Finally, it offers guidance on how to ensure that communication strategies cover all stages of the World Bank project cycle and the infrastructure project cycle.

In addition, this handbook is directed at reinforcing the following outcomes:

- Improving the quality of dam projects in the World Bank's own portfolio by enabling and facilitating functional partnerships that better reflect stakeholders' concerns about risk, empowerment, and opportunity
- Inspiring action to embed governance and anticorruption measures in dam projects through raising awareness of the key corruption vulnerabilities at all stages of the infrastructure cycle
- Drawing attention to the long-term capacity building of modern communication practices to promote a culture of communication in dam planning and management.

Origins of the Handbook

This handbook emerged from a discussion paper supported by the Development Communication Division (DevComm) of the World Bank that explored the notion of standards for communication in infrastructure development. Presented at the World Congress on Communication for Development in Rome in 2006, the paper received a positive reaction, and it then became World Bank Working Paper 121 (2007) in order to disseminate the concepts and information more widely.

That working paper outlines a systematic approach to developing voluntary standards for communication on infrastructure projects—standards linked to governance and sustainability themes. It recommends five follow-up initiatives relevant to dams. DevComm and the BNWPP collaborated on the first initiative, which is a direct response to the call by the World Bank Development Committee for infrastructure and communication practitioners to collaborate on improving communication strategies in all phases of the project cycle, focusing on dam planning and management themes. This handbook grew out of that response.

Acknowledgments

This handbook and its companion case studies in the World Bank working paper series benefited from the experience and advice of many people. Special thanks are extended to stakeholders on the Berg Water Project in the Western Cape Region of South Africa and the multiphase Lesotho Highlands Water Project in the Kingdom of Lesotho. And, in particular, we are grateful to the specialists from government, civil society, and the private sector who gave their valuable time to review drafts of the case studies and who offered their creative input. They are Liane Greeff, Jessica Hughes, Paul Roberts, and Charles Sellick, as well as the many representatives of the Department of Water Affairs and Forestry (DWAF), Trans-Caledon Tunnel Authority (TCTA), City of Cape Town, and Ninham Shand Inc. Thanks are extended as well to Leon Trump of the Lesotho Highlands Water Commission; Masilo Phakoe and Motulatsi Nkhasi at the Lesotho Highlands Development Authority; Johann Claassens and David Keyser at TCTA; DWAF staff, particularly Wille Croucamp and Reggie Tsekaseka; and Marcus Wishart, Rafik Hirji, and the late Dan Aronson of the World Bank.

Many development practitioners at the World Bank commented on the initial drafts of the handbook and offered creative input and support, including (in no particular order) Daryl Fields, Alessandro Palmieri, Sumir Lal, Adesinaola Michael Odugbemi, and Paul D. Mitchell.

Special thanks are extended to the copyeditor, Sabra Bissette Ledent, and to Mark Ingebretsen, Stuart Tucker, and Santiago Pombo of the World Bank's Office of the Publisher for their outstanding editorial and publishing support.

Finally, we would like to acknowledge the assistance and funding provided by the Bank–Netherlands Water Partnership Program for the preparation of this handbook, and we would like to express our thanks to the colleagues within the World Bank who facilitated the approval and funding, especially Daryl Fields, Alessandro Palmieri, Xiaokai Li, and Diego Juan Rodriguez.

Note

1. Lawrence J.M. Haas, Leonardo Mazzei, Donal T. O'Leary, and Nigel Rossouw, *Berg Water Project: Communications Practices for Governance and Sustainability Improvement*, World Bank Working Paper No. 199 (Washington, DC: World Bank, 2010); Lawrence J.M. Haas, Leonardo Mazzei, and Donal T. O'Leary, *Lesotho Highlands Water Project: Communications Practices for Governance and Sustainability Improvement*, World Bank Working Paper No. 200 (Washington, DC: World Bank, 2010).

About the Authors

Leonardo Mazzei is an international expert in strategic communication who assists governments on difficult reform programs and high-risk projects. As a senior communication officer in the Development Communications Division of the World Bank, he worked in Africa, Central and East Asia, Eastern Europe, and Central America with a focus on governance, infrastructure, and private public partnerships. He has extensive experience in political risk analysis, public consultations processes, conflict resolution, and opinion research. Currently, he leads communications for the Structured and Corporate Finance Department of the Inter-American Development Bank, supporting private sector investments in Latin America and the Caribbean.

For nearly 30 years, **Lawrence J. M. Haas** led multidisciplinary teams working with water and energy utilities and commissions in Africa and Asia to implement projects and capacity building. He was a team leader in the secretariat of the World Commission on Dams. Currently, he is an independent consultant in dams and sustainable development issues in the water and energy sectors for organizations such as the World Bank, World Wide Fund for Nature, International Union for Conservation of Nature, Asian Development Bank, and the Mekong River Commission Secretariat.

Donal T. O'Leary is the lead water adviser at Transparency International (TI) and is a cofounder of the Water Integrity Network. He was one of the principal authors of *The Global Corruption Report 2008: Corruption in the Water Sector* (Cambridge University Press) and represented TI on the Hydro Sustainability Assessment Forum, a multistakeholder platform that produced the Hydro Sustainability Assessment Protocol, published in November 2010. From 1982 to 2005 he was a senior power engineer with the World Bank, working primarily with the Energy and Water Groups for the South Asia and Africa Regions. He worked with Siemens AG under the World Bank Staff Exchange Program and represented Siemens in the Industry Group associated with the World Commission on Dams.

Abbreviations

BNWPP	Bank–Netherlands Water Partnership Program
BWP	Berg Water Project
CAS	country assistance strategy
CBA	communication-based assessment
CNA	communication needs assessment
CRA	corruption risk assessment
DC	Development Committee (World Bank)
DevComm	Development Communication Division (World Bank)
EFA	environmental flow assessment
EIA	environmental impact assessment
EMC	environmental monitoring committee
EMP	environmental management plan
Eskom	Electricity Supply Commission (South Africa)
GAC	Governance and Anticorruption Strategy (World Bank Group)
GIP	governance improvement plan
GIS	governance improvement strategy
G-S-C	governance-sustainability-communication linkage
IFR	instream flow requirements
IHA	International Hydropower Association
IP	integrity pact
ISA	integrated sustainability assessment
IWRM	integrated water resource management
LHWP	Lesotho Highlands Water Project
PM	project manager
PM&E	participatory monitoring and evaluation
PPP	public participation process
PRSP	poverty reduction strategy paper

SEA	strategic environmental assessment
TI	Transparency International
TTL	task team leader (World Bank)
WCD	World Commission on Dams
WWF	World Wide Fund for Nature (WWF International)

Introduction

Communication for development is a comparatively new field that offers new tools and techniques to support inclusive and informed decisions in the planning and management of large water and energy infrastructure projects, including dams. Rethinking the approach to communication on dam projects is also timely in today's policy context. A window of opportunity has opened to tie in governance reform (including fighting corruption), poverty reduction, and communication with today's challenges in sustainable infrastructure development. Progress on any one of these aspects requires effective communication with stakeholders and interests.

Presently, there are no standards for communication in the infrastructure project cycle, including that of dam projects. The World Bank's Development Committee recognized the significance of this gap in 2006 when it called for improving governance on infrastructure, specifically by developing communication strategies for all phases of the project cycle.[1] International civil society, industry, and donors endorsed the recommendation set out in the discussion paper "Strengthening Bank Group Engagement on Governance and Anticorruption," which was presented at the 2006 meeting of the Development Committee.[2]

From an operational perspective, communication can advance efforts to mainstream sustainability in hydropower development. Similarly, communication can advance efforts to place dam planning and management in an integrated water resource management context and reach a consensus on anticorruption measures.[3] Moving forward requires clarity on practical, effective ways to build a communication capacity within both project teams and implementing agencies and on the tools needed by donors, international nongovernmental organizations (INGOs), and others to help borrowers advance these aims at the sector and project levels.[4]

This handbook aims to help foster a "communication culture" that will accommodate the wide range of stakeholder interests in dam planning and management in ways appropriate to the development context of today and the need to promote solutions to sustainability challenges. It seeks to create awareness among practitioners of the benefits and costs of improving the role of communication in infrastructure development. It also demonstrates how communication helps to improve governments' capacities to address corruption issues in infrastructure. Finally, this handbook is aimed at building the capacity of project teams and government officials to effectively adopt and adapt modern communication principles and tools to cover all stages of the dam project cycle.

The Challenge

The provision of water infrastructure intersects with many different aspects of governance reform and democratization of development processes. In turn, these aspects bring many new issues, complexities, expectations, and voices to the table, which task managers must accommodate. Fundamentally, more communication capacity is needed to engage a wider range of stakeholders to support the kind of functional partnerships required to deliver sustainable solutions, integrate governance reforms, and cope with growing complexity. Improving project risk management is central. Indeed, today risk is defined more comprehensively than in the past, encompassing risks to stakeholder expectations and values associated with a failure to deliver sustainable performance, as well as the risk of corruption and the reputational risk of all partners.

The integration of better communication in dam projects will foster a philosophy of good governance and sustainable solutions that is consistent with the World Bank's Governance and Anticorruption Strategy and its leadership role. Corruption has manifest and disproportionate impacts on the costs and quality of service delivery to the poor, and otherwise undermines the infrastructure strategies of societies. Transparency and communication underpin most of the tools used to fight corruption. These tools include up-front awareness raising to overcome barriers to action and measures to prevent and detect corruption in collaborative approaches. Sustainability goes beyond the traditional notion of optimizing the design and performance of dams as physical assets for water and energy service provision. Today, it also encompasses the notion of optimizing the development performance of dams. Sustainable outcomes are delivered by innovative solutions that balance the economic, social, and environmental performances of dams within a context that can be understood and shared by stakeholders. Communication captures and fuses these elements, informs judgments, and links multiple actors to deliver sustainable performance.

The notion that communication plays a key role in introducing governance reforms and sustainable infrastructure provision is not new. What is new is linking these mutually reinforcing elements in dam planning and management as an explicit strategy. This new strategy makes conceptual and practical sense; it helps to manage risks and expectations systematically.

Conceptually, the opportunity for three-way synergy among communication, governance, and sustainability stems from typical situations in which, for example, reforms, whether for anticorruption or water governance, are set within a participatory framework, requiring more open and inclusive decision processes, or governance and sustainability connect in many issues and interests. These situations involve the same interested and affected parties. They inform decisions on competing interests, or they cooperate on shared interests. Synergy is achieved when the issues are tackled holistically, thereby gaining more for less. Communication supports anticorruption efforts, which in turn support sustainability by mitigating the risks to the social and environmental performance of dam projects and large water infrastructure more broadly.

Experience from the Field

For the people on the ground, many of these issues are inseparable. For example, when dam project authorities meet with host communities or river users to talk about one set of issues such as the environment management plan (EMP), those attending will generally raise all of the issues that concern them. Controversial issues are not easily dealt with in isolation—confidence and trust are built by resolving issues that affect the whole relationship.

This handbook builds on two case studies in order to identify current practice and to draw insights and lessons from the field: the Berg Water Project in South Africa (described in appendix A) and the Lesotho Highlands Water Project in Lesotho (described in appendix B). The Berg Water Project is a clear illustration of the synergy among issues that is important to stakeholder and project management. In that project, the multistakeholder environmental monitoring committee (EMC) became the main platform for dialogue on how to resolve the social sustainability concerns of the project, including the equitable sharing of benefits with the host community. In response to these concerns, the EMC set up working groups to deal with aspects of project governance and sustainability. EMC members also agreed on a comprehensive communication protocol. This protocol guided member interaction, communication with respective constituencies, and engagement of the media.

Where to Start

This handbook offers five "how-to" communication-based steps for improving governance and sustainability in dam planning and management. These steps are intended to increase transparency and accountability in the cooperation among government, civil society, and the private sector in structured processes; integrate sustainability approaches and thinking; and induce recognition that different aspects (environmental, social, financial, and institutional) of sustainability are intertwined. The five how-to steps can be summarized as follows:

1. **Start with a communication-based assessment (CBA).** The communication-based assessment is the first opportunity to identify how communication can improve project design as well as governance and sustainable performance. In fact, this assessment is the first step in the World Bank's four-phase development communication methodology, described in chapter 2. The CBA, which can vary in depth and duration based on needs, helps to identify knowledge gaps, stakeholders' present perceptions, communication systems, problems and risks, and possible solutions.

2. **Benchmark against good practice.** Benchmarking plays an important role in the preparation of a communication strategy. Generic benchmarking steps include the following:

   • **Measure and then manage stakeholder expectations.** Managing expectations requires understanding what all stakeholders expect of the project and what they fear to lose, and then managing the risks people perceive as important. Communication-based assessment tools are available to task managers seeking to detect and avoid the controversies fueled when the expectations of one or more stakeholder interests is not recognized from the start, or is not accommodated in the project design, or is simply not viewed as "important" in the scheme of things as the project proceeds. The assessment equips project managers with "radar" to identify potential problems early and manage them better.

   • **Empower the relevant voices.** Success in enhancing governance and sustainability hinges on effective coalitions of government, private sector, and civil society actors. Using two-way communications to give people a voice in development decisions that affect them improves the development performance of dam projects and allows people to participate in a meaningful way in multistakeholder partnerships. Empowering the relevant voices also helps project teams to manage and resolve the trade-offs on different aspects of sustainable performance. Decision-making authorities at all levels are better informed in negotiating trade-offs when all voices are empowered.

   • **Build trust for functional partnerships.** Establishing a policy of clear two-way communications from the start is essential to creating, building, and maintaining trust—not only stakeholders' trust in authorities to act in "good faith" but also their trust in the mechanisms that will reconcile their interests. Trust is achieved through transparency, commitment to participation, culturally appropriate communication and feedback, and written commitments that can be verified and upheld. Trust is key to creating the space for agreement on sharing costs and benefits (within the policy framework and regulation) and on pursuing the adaptive management of dams. Trust between riparian states is of central importance to mutually beneficial cooperation on international rivers.

- **Add value for different groups of stakeholders.** Attention to value added is critical for host communities and affected river users, who typically live in remote rural areas. Value in this context is what these communities and users perceive to be and define as value—and it is in addition to the value added for the primary beneficiaries of water and energy services, who typically reside in towns and urban centers. Communication must underpin inclusive decision making on dam design, delivery, and performance to ensure that these factors add value for stakeholders. Benefits must be shared equitably and perceived as such.

- **Advocate forms of risk reduction.** Communication must support the identification and management of all forms of risk, with risk defined in an encompassing way, not narrowly. Communication can support a more holistic and integrated approach to risk assessment and management. In practice, it means ensuring that the risk exposures important to relevant stakeholder constituencies are identified, properly assessed, and reasonably addressed from the outset of a project. Paying attention to all forms of risks includes heeding all dimensions of sustainable performance. In addition, risk at the local level includes taking into account environmental risks and focusing not just on the voluntary risk takers, such as contractors, owners, and financing agencies, but also on communities.

3. **Formulate comprehensive communication strategies.** Each major project activity and partnership group would have a detailed communication strategy. Typically, a communication strategy for large water infrastructure would include four distinct dimensions: (1) development communication, which focuses on the design and supervision of public communication programs and identifies risks; (2) corporate communication, mostly for external audiences, which explains how the corporation reflects government policy and how it will implement reforms on the dam project in which it is involved; (3) advocacy, which defines messages to raise the profile of an issue and influence change; and (4) internal communication, an essential part of internal governance, which helps to enhance synergy and achieve internal consensus on goals and messages before going "outside." These dimensions will support public participation processes to inform choices on dam and non-dam options, communication with the people affected by a project to ensure that their concerns are reflected in project design, internal communication among institutional partners on implementation arrangements and coherence, coordination with sector entities to optimize the benefits of multipurpose dams, engagement with

end-user groups to locate key decisions about dams in the basin development context, and advocacy by government and water institutions to build support for the project.

4. **Improve and adapt communication along the project cycle.** Adoption of a greater emphasis on analysis to monitor stages dynamically shifts classical thinking on the analytical role of communication—or rather places more emphasis on analysis in monitoring stages dynamically linked to consultation and public participation processes. Communication strategies must be adapted for each stage of the infrastructure project cycle: planning, preparation, implementation, operation, and evaluation. For most projects, communication strategies must be updated to take into account (1) the changes in issues and risks between project stages, (2) the changes in the roles of the stakeholders, and (3) the typically long period of time between stages. In other words, the communication strategy must inform and support key decisions at each stage of the project cycle as actors, risks, and issues change.

5. **Invest in communication for governance and sustainability.** Improvements in communication must deliver value for the money and value for all stakeholders. Stakeholder value is achieved by investing in effective partnerships that also facilitate public acceptance, while reducing conflict that would otherwise increase project costs.

Implementation along the Project Cycle

At the *macro policy stage,* communication should support legislative and macro policy reforms by creating the strong culture of communication needed to promote water governance reforms from the top. At this level, communication strategies are aimed at building support for reforms, facilitating the kinds of fundamental behavioral shifts needed for water reforms that go to the heart of attitudes about water, and helping to manage the tension between local interests and national imperatives or national policy achievements. That ethos must be incorporated into how the public and private enterprises responsible for implementing reforms and infrastructure provision actually behave, and how corporate communication strategies reflect these behavioral changes. The relevant stakeholders include legislators, politicians, and representatives of the government, civil society, media, and private sector interested in the political economy of water and development.

At the *strategic planning stage*—that is, the sector- and basin-level strategic planning processes that lead to the identification of programs and projects—communication should reinforce the political legitimacy of decisions. It does so by explaining the rationale for strategic studies and comprehensive options assessments in order to reinforce the political legitimacy of decisions reached, while also clearly describing the convergent and divergent views and how trade-offs were reached. Similarly, communication should help clarify the infrastructure strategy on which decisions for a particular project are based. It would explain how the strategy was derived from basin-level water resources planning, and especially how it affects the water services received by different consumer groups, highlighting the impacts of the expansion of services to low-income groups and the impacts of tariffs. Communication strategies for public participation processes must reveal the different entry points for public input and otherwise seek to maximize opportunity, voice, and participation. They must then ensure that the government responds to the results of public participation so that trust in the outcome and future participation is not jeopardized. The relevant stakeholders include representatives of the various interested and affected parties and water institutions in the river basin.

At the *project preparation stage*—that is, the project-level studies that confirm the design of infrastructure projects up to approval and financing—those undertaking communication and social analysis need first to understand how different groups actually see the project—their attitudes, perceptions, anxieties, and expectations—and thereafter what each group of interested and affected parties requires in order to facilitate two-way communications. To establish trust and add value, communication efforts should facilitate the early meaningful involvement of interested and affected parties by detailing progress, sharing information, involving stakeholders in knowledge management activities, and addressing issues relevant to the parties. The relevant stakeholders again include basin residents and communities receiving water and energy services from the project, but at this stage more emphasis is placed on the community hosting the project.

At the *project implementation stage*—that is, after approval and through construction, implementation, and commissioning—communication should be directed at maintaining trust, sustaining dialogue, and expanding partnerships, thereby demonstrating the importance of effective dialogue mechanisms. Communication should also be directed at increasing the transparency of key decisions on project preparation, design, and operating strategies by using open, participatory mechanisms and taking the preliminary steps needed to reduce the corruption risks in procurement contracts. Successful communication at this stage helps set the scene for a partnership in the operating phase. The relevant stakeholders are the same as those involved in the previous stage, but now also include the suppliers of goods and services, contractors, and construction workers involved in the project.

At the *project operation stage*, communication is related to the sustainable integration of the project within the culture and economy of the host community. Similarly, communication revolves around integration of the operation within the evolving catchment management strategy. Transparent and effective communication should empower a dialogue among the interested and affected parties about compliance monitoring processes and adaptive management of the facilities and of downstream releases from the reservoir within the framework of the regulations. Communication strategies also must address the need to build a communication capacity in new water institutions. The relevant stakeholders include the host community, basin communities, water institutions, and consumers, as well as regulatory oversight bodies.

At the *project evaluation stage,* communication resources, processes, and outcomes would be assessed against objectives, critically considering the effectiveness of the communication strategy and the extent to which project communication added value to outcomes for the resources provided. Monitoring and evaluation should include a yearly report on the performance of project communication. Similarly, evaluators should try to capture convergent and divergent views, including the views of all stakeholders and project partners, and highlight successes and failures that could help improve communication on other projects. Where possible, they could use a multimedia approach to reflecting on the development process in the public sphere.

The budgetary implications of introducing good communication practices in support of governance and sustainability improvements in dam projects depend on a project's complexity. For example, spending on communication activities in the Lesotho Highlands Water Project (LHWP) described in appendix B increased from the early Phase IA and IB project preparation stages to the current Phase II. Communication was a very small component of the LHWP's total budget in the 1990s and early 2000s, during the Phase IA and IB project preparation works, but by 2008 communication and public stakeholder consultation activities related to the Phase II feasibility studies were accounting for more than 35 percent of the total budget. The Nam Theun 2 project in Laos budgeted $250,000[5] to contract out the CBA activities and collaborative preparation of the communication strategy for the implementation phase. This amount did not include the supervision costs, and it was one element of the total project communication budget. The Bumbuna project in Sierra Leone budgeted $200,000 for the CBA, communication action plan development during project preparation, and two years of implementation.

Organization of This Handbook: Multiple Entry Points

This handbook consists of three chapters, each designed to enable practitioners to use any chapter independently, depending on their needs and interests. The first two chapters are in modular formats for clarity and ease of use.

Chapter 1 explains why modernizing communication in dam planning and management is in everyone's best interests and why linking governance, sustainability, and communication adds value for all stakeholders.

Chapter 2 offers guidance for task managers who wish to adopt communication practices that help to advance governance and sustainability themes in dam planning and management. The guidance is aimed at improving communication at different stages of the project cycle, as advocated by the World Bank's Development Committee.

Chapter 3 is a primer on the communication, anticorruption, and sustainability tools and techniques that can be applied to dam planning and management by the infrastructure practitioners and task managers supervising the preparation and implementation of activities throughout the project cycle.

A handbook adds value only to the extent that it stimulates thinking and offers relevant guidance. Figure 1 and the explanation that follows suggest six ways in which World Bank task team leaders (TTLs), country-based project managers (PMs), and other practitioners can get the most from this handbook and its companion products:

- **Benchmark to identify room for improvement.** First and foremost, use this handbook to benchmark against criteria that reflect good practice (e.g., to identify gaps or missed opportunities) and to assess ways to add value in exploiting synergy among governance, sustainability, and communication.
- **Facilitate task team discussions.** Use the handbook to stimulate the exchange of new ideas and thinking among members of task teams or their counterparts and to structure one-on-one discussions between communication specialists and task team area specialists to serve as input for communication needs assessments and communication strategies.

Figure 1. How to Get the Most from This Handbook: Entry Points

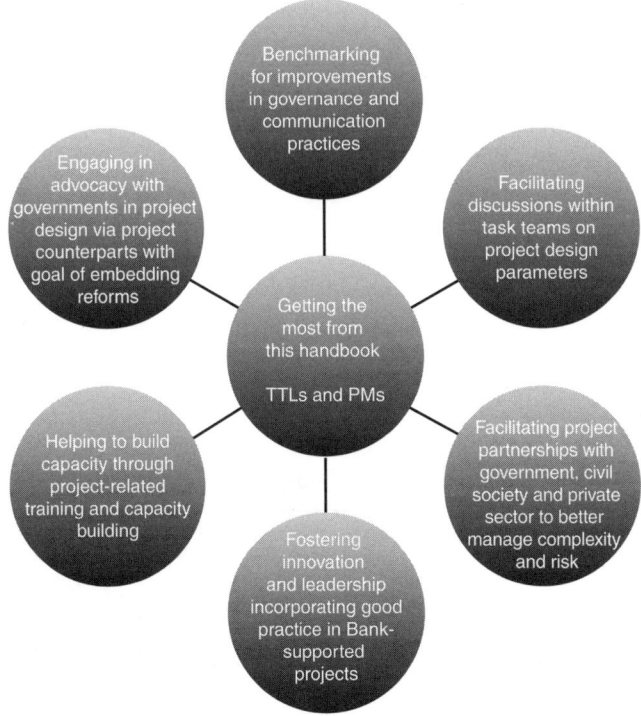

Source: Lawrence. J. M. Haas.

- **Engage in advocacy with government counterparts on project design.** Use the handbook as a basis for discussion between task teams and government counterparts on the steps and technical advice available for decision makers on the value of embedding good practice in World Bank-supported projects. This discussion is consistent with the Bank's broader advocacy role.
- **Facilitate project partnerships.** Circulate the handbook among stakeholders from government, civil society, and the private sector in World Bank-supported projects to help facilitate the formation and effective functioning of partnerships and to clarify thinking and priorities on what communication support adds the most value.
- **Foster innovation and leadership in the demonstration of good practice.** Use the handbook in suggesting good practice on World Bank-supported projects as a model for sector, country, regional, and wider knowledge sharing.
- **Help build capacity within project teams**. Offer information on training to incorporate in World Bank-supported projects.

Notes

1. See World Bank, "Strengthening Bank Group Engagement on Governance and Anticorruption" (discussion paper, meeting of Development Committee, Washington, DC, July 2006).
2. Referring to infrastructure more generally, the discussion paper (ibid.) noted:

   Because of the importance of an effective dialogue on issues of fraud and corruption, it is important to develop an effective communications strategy that covers all phases of the project. The communications plan must provide for consistent messages to be conveyed to all of the relevant stakeholders: government officials in the implementing agency; contractors, suppliers, and consultants who may be involved in bidding on the project; members of civil society affected by the project; and (as appropriate) the local press. The role of the media may be especially important if the plan includes the use of publicity—both positive and negative stories—as a tool for reducing the level of fraud and corruption in Bank projects. The objective would be to highlight both noteworthy achievements in quality, cost-effectiveness, and sustainability, as well as any incidents of alleged collusion, fraud, or corruption.

3. In today's context, partnership approaches are pursued to handle the risks and complexities of sustainable infrastructure development and management, and many new stakeholder interests are involved. This situation not only elevates the need for communications but also creates opportunities for synergy.
4. See also Larry Haas, Leonardo Mazzei, and Donal O'Leary, *Setting Standards for Communications and Governance: The Example of Infrastructure Projects* (Washington, DC: World Bank, 2007).
5. All dollar amounts are U.S. dollars unless otherwise indicated.

Advocacy and Awareness Raising

Five key messages on advocacy and awareness raising

1. **Communication.** Many voices, from dam-affected communities to the World Bank's Development Committee, have been raised about the need to strengthen and modernize communication practices in dam planning and management. The knowledge and tools are available to create a "culture of communication" on dams that is demand-responsive, is practical, and adds value for all stakeholders.

2. **Governance and anticorruption.** Infrastructure is classified as a sector at high risk of corruption. Corruption has manifest and disproportionate impacts on the cost and quality of service delivery to the poor, and otherwise undermines infrastructure strategies of societies. Transparency and communication underpin most tools to fight corruption. These tools include up-front awareness raising to overcome barriers to action and measures to prevent and detect corruption in collaborative approaches.

3. **Sustainability.** Sustainability goes beyond the traditional notion of optimizing the design and performance of dams as physical assets for the provision of water and energy services. It also encompasses the notion of optimizing the development performance of dams. Sustainable outcomes are delivered by innovative solutions that balance the economic, social, and environmental performance of dams in a context that can be understood by stakeholders. Communication captures and fuses local and expert knowledge, informs judgments, and links multiple actors to deliver sustainable performance.

4. **Partnership approaches.** Communication capacity forms and activates partnerships at all levels. It is essential to genuinely involve affected people in decisions on dams and to capture the synergy of pro-environment, pro-poor, and anticorruption concerns in dam planning and management (see number 1).

5. **Project management.** Investing in communication is a practical response to the challenges of managing stakeholder expectations across an expanding range of issues and growing complexity. Communication is integral to risk management, informing judgments about the risk of corruption and the risks to sustainable performance perceived as important to stakeholders. Investing in communication is part of the philosophy of "doing more for less"; it is a central component of risk management and mitigation.

Bridging the Communication and Infrastructure Paradigms

Two broad views of communication on large dam projects prevail. The first is that communication, like transportation, is a derived demand—that is, a means to an end.[1] In the context of dam planning and management, the demand for communication derives from the need to enhance sustainable performance, which is defined in more encompassing ways today than in the past, optimizing the wider role of dams in development.

The second view of communication is as a professional discipline in its own right, uniquely positioned to offer advice and new tools to help task managers advance pro-poor, pro-environment, and anticorruption solutions in the provision of infrastructure.

Understanding the interplay between these two views is essential to improvements in communication practices in the over 140 countries that are operating large dams and may plan to build more. The starting point is collaboration between infrastructure and communication practitioners to identify contextual improvements in communication that add value for all stakeholders.[2] Some observations might shed light on the possibilities.

Global Investment in Water Infrastructure Is Set to Increase Significantly

In response to the multiple dynamic pressures on their water resource systems, most developing countries are adjusting their infrastructure priorities and placing economic growth on inclusive and sustainable pathways where populations are growing:

- Current estimates of financing needs for infrastructure development and for operation and maintenance across all sectors in developing countries stand at about 7 percent of the gross domestic product (GDP), and they range as high as 9 percent of GDP in low-income countries.[3] Water investments, including hydropower and water and sanitation, represent a significant portion of the total in many countries.
- A quarter of the world's people (1.69 billion) are living in areas characterized by physical water scarcity. A further billion are facing economic water scarcity because of the lack of investment in water or the human capacity to keep up with demand.[4] As water scarcity grows, storage dams are receiving more attention as part of the "options mix" in the imperative to reconcile water demand and supply within river basins.

"By promoting economic growth strategies based on expanded infrastructure which are environmentally responsible and socially acceptable we are bringing a sustainable future closer to today's reality."

—Katherine Sierra, Vice President for
Sustainable Development, World Bank (2007)

- Similarly, multipurpose dam projects are receiving more attention as a strategy to maximize development returns on infrastructure investments. In international rivers, multipurpose development expands the scope for mutually beneficial sharing across a wider range of benefits, looking beyond the narrow view of water as a finite commodity to be divided. The new generation of international agreements reflects these aspects.
- Today, both the public and private actors are placing more emphasis on the potential contribution of dams to meeting the interrelated policy priorities of energy security, water security, and multicountry regional economic integration.[5] Other drivers coming to the forefront are climate change mitigation and adaptation.[6]

It is not only the new demands that are driving water infrastructure investment trends. Worldwide, the annual rate of loss in surface water storage due to reservoir sedimentation is estimated to be 0.5–1 percent of the total existing storage capacity.[7] Many countries are seeking ways to replace the lost strategic storage that threatens the existing capacity to deliver water and energy services as soil erosion and reservoir sedimentation rates accelerate.

In 2003 the World Bank began to scale up its involvement in hydropower, multipurpose dams, and other water infrastructure investment in recognition of global trends. By 2008 the Bank was the largest single external financier in the global water sector.[8]

> For dams and development, "the challenge is to find ways of sharing water resources equitably and sustainably—ways that meet the needs of all people as well as those of the environment and economic development . . . needs that are all intertwined."
>
> —World Commission on Dams (2000)

Water Infrastructure Is at the Confluence of Many Development Reforms
Along with the resurgence of investment in water infrastructure, many different strands of governance reform converge in dam planning and management. This convergence widens the range of stakeholder interests in decision processes at all stages of the infrastructure project cycle:

- Governance and anticorruption reforms in infrastructure, water, and energy services, where infrastructure is classified as a high-risk sector, if not the highest-risk sector
- Water governance reforms structured around a broader rethinking of water tariffs, water demand–supply reconciliation, and reform of water-related institutions
- Reforms in the management of freshwater resources based on integrated water resource management (IWRM) principles, including innovative

measures and incentive mechanisms to adapt land-water resource systems to climate change

- Reforms in project financing and public–private sector roles, all part of the dynamic shaping infrastructure strategies in most countries
- Assurance that the alleviation of poverty is an explicit focus in the provision of infrastructure, also related to meeting Millennium Development Goal (MDG) targets.

The cumulative effect of these governance transitions and their interrelated reforms that aim to deliver sustainable infrastructure solutions is more emphasis on balancing the social, economic, and environmental performance of dams.[9]

Dams as Communication-Intensive Projects

From a communication standpoint, modern forms of dam planning and management call for open, transparent, and inclusive processes that involve dialogue among stakeholders, as well as media relations and partnership approaches. In practice, decisions at all stages of the project cycle would benefit from improvements in communication such as:

- Public participation processes to reconcile demand and supply strategies and to inform the choice of dams and non-dam options
- Communication with the local project-affected populations to ensure that the project design and implementation reflect their needs and concerns, thereby building local knowledge
- Communication with key institutional partners on implementation arrangements
- Coordination with different sector entities to optimize the benefits of multi-purpose dams—for example, for power, flood management, irrigation, and water supply entities—during the project design and operation stages
- Engagements with IWRM entities and end-user groups to locate important decisions about the development and management of dams in the basin context and to reconcile the competing needs of all consumptive and non-consumptive water users
- Internal communication among and within the major institutional partners when various departments are involved in the project
- Advocacy, media relations, and public communication from governments and water institutions to engender support for the project.

"The development of infrastructure in the past has not seriously considered a systems or holistic approach.

There is a need for institutional integration among different stakeholders and sectors both horizontally and vertically including public-private partnerships."

—Sustainable Infrastructure in Asia: Overview and Proceedings, Economic and Social Commission for Asia and the Pacific (2007)

Communication analysis is especially important where there are deep-seated views. Reconciliation of multistakeholder partnerships and behavioral change is essential.[10] For example, lowering the walls between public and private investment embodies a shift in the rights and responsibilities of the government, civil society, and private sectors, and it requires new thinking (e.g., public– and community–private sector partnerships). Information sharing, dialogue, and negotiation are central to forming partnerships and to success.

Communication capacity is also critical for project innovation and the introduction of new dimensions such as benefit sharing, ensuring local communities become development partners in projects, and crystallizing the definition of *sustainable infrastructure*.

Adopting a New Generation of Tools

Communication tools are evolving. Chapter 3 offers thumbnail sketches of the range of tools practitioners can apply to improve project outcomes and better integrate governance and anticorruption (GAC), sustainability, and communication practices in dam planning and management.

Today's shift in emphasis defines emerging good practice around three dimensions: (1) governance and anticorruption; (2) provision of sustainable infrastructure; and (3) communication. These three dimensions form the foundation for good practice as infrastructure projects expand their purpose to meet wider development goals.

> "Corruption is a major cause of poverty as well as a barrier to overcoming it."
>
> —Peter Eigen, President, Transparency International

Governance and Anticorruption Orientations

Better transparency and accountability enhance the fight against corruption. Active cooperation among the government, civil society, and private sector helps to detect and prevent corrupt practices. Transparency International (TI), whose *Global Corruption Report 2008* focused on "Corruption in the Water Sector," highlights how manifest forms of corruption affect the poor and the environment disproportionately and undermine the infrastructure strategies of countries.[11]

Table 1.1, adapted from TI's analysis, illustrates the kinds of corruption vulnerabilities that exist in hydropower development, starting with the up-front identification of projects through the implementation, operation, and rehabilitation stages. These risk exposures are present in many large infrastructure ventures in the water sector.[12]

Table 1.1. Corruption Exposures in Hydropower along the Infrastructure Project Cycle

Options selection	Project planning	Contracting/bid evaluation	Construction/ implementation	Operation and rehabilitation
		Project Cycle ———→		
Limited options considered	Unnecessary studies	Nontransparent prequalification	Concealing substandard work	Commitments not kept
Predetermined, biased	Technical specifications biased to a particular technology	Confusing tender documents	Agreeing to unwarranted contract variations	Underfunding of environmental and social mitigation obligations, no money plea
Nontransparent	Over or under design	Nontransparent or nonobjective selection procedures	Creating artificial claims	Corruption in O&M procurements
Limited public involvement	Poor EIA	Bid clarifications not shared with other bidders	Biased project supervision	Insurance fraud on equipment and performance guarantees
Undue influence by proponents	Nontransparent EIA clearance	Award decisions not made public, or not justified	Bribery to avoid project delay penalties	Cycle of corruption vulnerability repeated in rehabilitation works
Selection of unnecessary projects		Deception and collusion Agents' fees and subcontracts to hide bribes or kickbacks	Corruption in resettlement and compensation A blind eye to construction environmental violations	

Source: Donal T. O' Leary.
Note: EIA = environmental impact assessment; O&M = operations and maintenance.

A comprehensive approach to tackling corruption in infrastructure such as dams is crucial at all levels: national, sector, and project. Both the demand and the supply dimensions of corruption require attention. The supply-side dimension encompasses North-South flows of money, equipment, and services. Countries must cooperate on the enforcement of international conventions such as the Organisation for Economic Co-operation and Development's Convention on Combating Bribery of Foreign Public Officials.[13]

But good governance and anticorruption measures are not just about laws, sanctions, and prosecutions.[14] Stamping out corruption requires a fundamentally collaborative approach in which affected people are given the chance to speak out and share their concerns.

Figure 1.1 illustrates the range of actors that broadly compose the public sector governance system around infrastructure. This figure appeared in the 2007 World Bank Development Committee (DC) strategy paper "Strengthening World Bank Group Engagement on Governance and Anticorruption."[15]

Figure 1.1. Stakeholders in the Public Sector Governance System for Infrastructure

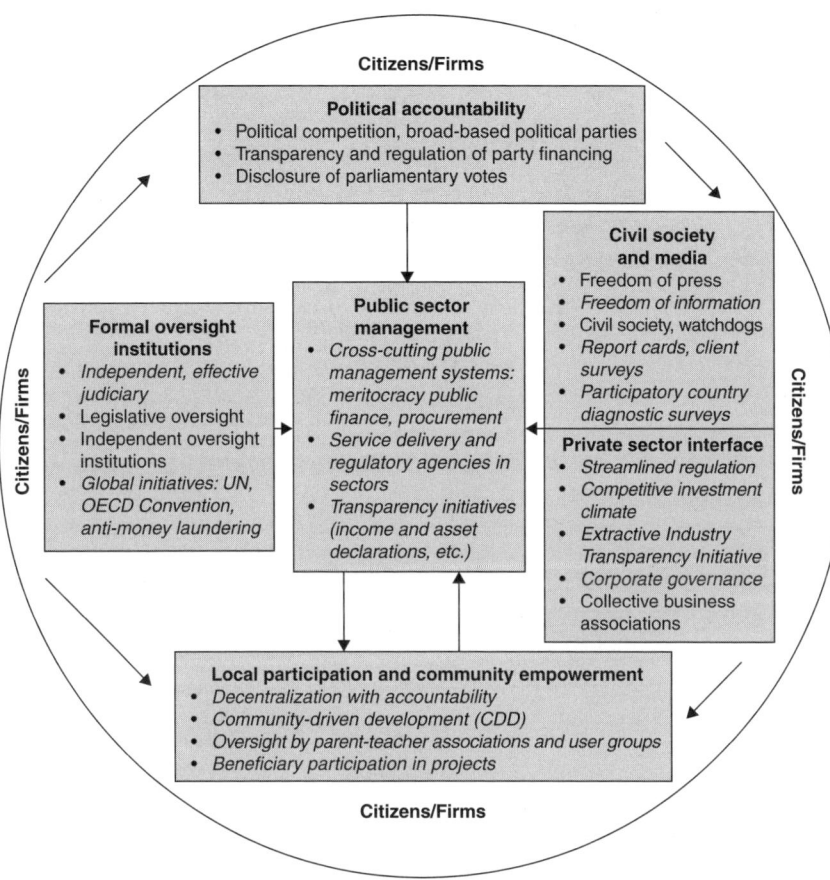

Italics = Areas supported in varying degrees by WBG operations.
Source: World Bank, "Strengthening World Bank Group Engagement on Governance and Anticorruption," Strategy Paper, Development Committee, Washington, DC, 2007.

Good practice in the prevention and detection of corruption in dam planning and management encompasses the actions of many institutions and people (figure 1.1) who must act within their own domains while participating in partnerships at different levels. These institutions and people include:

- Government executive and cross-cutting control agencies responsible for public finance and human resource management, and frontline regulatory and service provision agencies
- Formal oversight institutions outside the executive, including the judiciary, legislature, and other independent oversight institutions
- Subnational government organizations and local communities with their own infrastructure and service provision responsibilities, and often with their own local arrangements for control and accountability
- Political actors and institutions at the apex who are setting the broad goals and direction

- The dam industry, comprising equipment suppliers, contractors, consultants, and financial intermediaries
- Civil society and the private sector in their roles as both watchdogs with the independent media and the "recipients" of services and regulations and thus a potential source of advocacy and pressure for better performance.

Chapter 3 describes the practical tools that task managers can use to address corruption risks at different stages of the infrastructure cycle. The common theme of these tools is their underpinning of transparency and communication, from up-front awareness-raising and advocacy activities, to efforts to overcome barriers to action, to the mechanisms to identify, agree on, and implement priority measures in collaborative ways.

What are the differences between past GAC practices and a communication perspective? As yet there is no standard or required approach to embedding anticorruption measures in dam projects. On Bank-supported projects there are, of course, the protections afforded by disclosure policies and other prescriptions for good governance that relate to transparency, accountability, and formalized audits. However, as TI notes in its comments on the World Bank Governance and Anticorruption Strategy, these measures are necessary but not sufficient.[16]

As discussed further in chapter 3, concerted action to manage corruption risks on dam projects starts with two practical approaches:

1. Signal the need for an anticorruption component in the up-front identification and conceptualization of a project,[17] followed by an in-depth corruption risk assessment (CRA) in project preparation.
2. Embed a governance improvement plan (GIP) in the project governance framework as a routine element of the risk management system in the implementation stage, and update the GIP at the start of operations and periodically thereafter.

The GIP would prioritize measures to mitigate risks identified in the corruption risk assessment. A communication-based assessment (CBA) would inform judgments on what is needed for effective collaboration at each step. The corruption risk assessment also should link to the communication needs assessment (CNA) in order to prioritize communication support on each measure.

Concerted action in combating corruption means maintaining pressure to build a multistakeholder consensus in the public sector governance system around infrastructure. This finding is illustrated in figure 1.1.

- Tackling corruption is in the interests of all stakeholders because it affects the level and quality of services, the equitable distribution of benefits, and sustainability.
- Tackling corruption is a gradual, long-term process that involves changing attitudes that tolerate corrupt practices, retraining staff, and restructuring institutions.

- Concerted action requires assigning priority to the methods for achieving transparency, engaging civil society, assessing mechanisms for regular independent monitoring, and enlisting the support of stakeholders.

Table 1.2 illustrates the differences in enabling or moving from past or current practice to good practice. Perhaps the most important measure of success is that corruption risk assessment and mitigation are routinely a part of a comprehensive, integrated risk management framework on all dam projects.

Table 1.2. What Is the Difference between Thinking and Practice? Preventing and Detecting Corruption in Dam Planning and Management

Typical past thinking or practice ──────────▶	Enabling or representing good practice
Change in thinking	
✄ Corruption is always a concern, but it is not the same priority as getting water and energy services flowing.	✓ Combating corruption is in the interest of all stakeholders. In the end, governments and citizens will pay a price: lower service levels, investments, and incomes.
✄ Government is responsible for governance and anticorruption measures and the appropriate legislation.	✓ Fighting corruption is a shared responsibility. Coalitions are required to fight corruption at each stage of the infrastructure project cycle.
✄ Corruption primarily occurs in construction and the procurement of goods and services. "Grand corruption" is the most important.	✓ "Grand" and "petty" forms of corruption have disproportionate impacts on the poor. Corruption risks exist at all stages of the project cycle.
✄ Focus is on detecting corruption.	✓ Focus is on prevention and detection, supported by the government's prosecution powers.
✄ Corruption is not part of the normal project risk assessment; it is largely related to reputational risk.	✓ Dealing with corruption risks is an integral part of comprehensive risk management. Corruption threatens all dimensions—economic, social, and environmental—of the sustainable performance of dams.
Change in practice	
☹ No corruption risk assessment (CRA) is undertaken for dam projects.	☺ CRA looks at corruption risk exposures at each stage of the project cycle as part of an integrated risk assessment framework, with advocacy used to make the case.
☹ No formal program identifies or manages corruption risks, apart from following normal procedures, including procurement and disclosure.	☺ Governance improvement plan (GIP) is incorporated in projects to reflect risks identified in a CRA, covering the design, implementation, and operation stages.
☹ Communication is not explicitly deployed or emphasized as a tool to help combat corruption.	☺ Communication strategy is prepared to support the GIP, using advocacy, awareness raising, and support of multistakeholder partnerships for each measure.
☹ Civil society, local communities, and consumers of services have no formal role in detecting and preventing corruption.	☺ Local stakeholders fight corruption through monitoring, watchdog action, Citizen Report Cards, and hotlines.
☹ No lessons are drawn from country experience combating corruption on other dam projects or in other infrastructure sectors.	☺ The CRA and GIP are benchmarked against emerging national, regional, and global practices, and this information is shared with stakeholders while formulating the GIP.

Provision of Sustainable Infrastructure

Sustainable development in the context of today's dam projects means balancing the economic, social, and environmental performance of dams without compromising the ability of future generations to adjust this balance over time.[18]

Decisions on this balance must be informed by stakeholder values and expectations across a range of development issues and themes. Equally important, the prospect of progress across generations calls for building in, whenever feasible, flexibility to allow adaptive management of both the physical structures and the operating procedures of new dams and dam rehabilitation works.[19]

Figure 1.2 illustrates generic governance issues and themes at the forefront of dam planning and management when looking through a multifaceted sustainability lens.[20] It shows examples of process-oriented tools, such as environmental assessment, and performance-oriented tools, such as benefit sharing.

At the center of figure 1.2 is communication for sustainable infrastructure. This configuration illustrates the notion of interdependence and the need to connect to stakeholder interests around each theme.

As illustrated, two-way communication empowers inclusive decision making and multistakeholder partnerships, which are often the best mechanisms for delivering sustainable performance on dams when competing interests must be reconciled.

Chapter 3 offers thumbnail sketches of tools to support each sustainability theme in figure 1.2. These sketches note the type of good practice communication support available for each tool.

Figure 1.2. Old and New Sustainability Themes in Dam Planning and Management

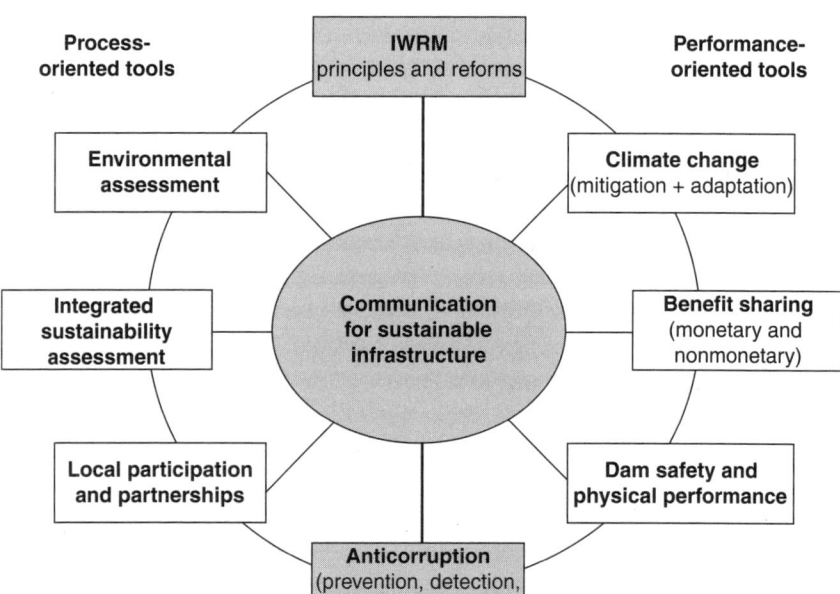

Source: Lawrence J. M. Haas.

Box 1.1.   Featuring Communications and Governance in the Hydro Sustainability Assessment Protocol (HSAP)

"Sustainability [in hydropower] is a fundamental component of social responsibility, sound business practice and natural resource management. . . . [I]t is a collective responsibility of government, business, civil society consumers and individuals."

> —International Hydropower Association, Sustainability Assessment Guideline (2004) and Sustainability Assessment Protocol (2006)

The Hydro Sustainability Assessment Protocol (HSAP),[a] issued in November 2010, is a sustainability assessment framework for hydropower development and operation. It facilitates the production of a sustainability profile for a hydroelectric project for a number of priority sustainability topics.

The Protocol includes four stand-alone sections that reflect different stages of hydropower development. Through an evaluation of basic and advanced expectations, the Early Stage tool may be used for risk assessment and for dialogue prior to advancing into detailed planning. The remaining three documents, Preparation, Implementation, and Operation, set out a graded spectrum of practice calibrated against statements of basic good practice and proven best practice. The graded performance within each sustainability topic also provides the opportunity to promote structured, continuous improvement.

Each of the Protocol sections relating to Preparation, Implementation, and Operation include sustainability topics entitled "Communications and Consultation" and "Governance"; in addition, the Preparation and Implementation sections of the Protocol contain a section on "Procurement." The Communications and Consultation topic addresses "the identification and engagement with project stakeholders, both within the company as well as with external stakeholders (e.g. affected communities, governments, key institutions, partners, contractors, catchment residents, etc.)" with the intent of establishing "a foundation for good stakeholder relations throughout the project life. The Governance topic "addresses corporate and external governance considerations for the project [with] the intent that the developer has sound corporate business structures, policies and practices; addresses transparency, integrity and accountability issues; can manage external governance issues (e.g. institutional capacity shortfalls, political risks including transboundary issues, public sector corruption risks); and can ensure compliance." The Procurement topic "addresses all project-related procurement (works, goods and services) with the intent that procurement processes are equitable, transparent and accountable; support achievement of project timeline, quality and budgetary milestones; support developer and contractor environmental, social and ethical performance; and promote opportunities for local industries."

The Protocol was the result of a two-and-a-half-year process under the guidance of the Hydropower Sustainability Assessment Forum (HSAF), which consisted of representatives of developing countries (China and Zambia), developed countries (Germany [observer], Iceland, and Norway), the finance sector (Equator Principles Financial Institutions and the World Bank [ob-

(continued next page)

server]), the hydropower sector (the International Hydropower Association and HydroTasmania), and nongovernmental organizations (Oxfam, The Nature Conservancy, Transparency International, and World Wide Fund for Nature).

a. http://www.hydropower.org/sustainable_hydropower/HSAF_Hydropower_Sustainability_Assessment_Protocol.html.

What are the differences between past practices and a communication perspective? There is a continual process of knowledge development, innovation, and improvement in the tools available for dam-related planning, management, and technical issues, all of which enhance the sustainable performance of dams. Many countries are experiencing small but frequent improvements in practices over time. The accumulation of these improvements amounts to a radical change in thinking about dams and sustainability.

Take, for example, environment flows. In less than a decade environmental flow assessment (EFA) and provision has gone from a relatively obscure practice to one now widely recognized, even if adoption is not yet as widespread.[21]

Today, there is a broad call for integrated sustainability assessment (ISA) approaches and better integration of existing sustainability assessments, recognizing the intertwining of different sustainability aspects: environmental, social, financial, and institutional.

It is not enough to look at one aspect in isolation, such as social sustainability, without considering whether measures to ensure it are sustainably financed, or whether the institutional capacities for monitoring, compliance, and communication that empower people to take local action have sufficient resources.

The call for ISA approaches ties into that of working to mitigate risks that stakeholders perceive as the most important. The underlying trend is to define risk comprehensively—that is, as risk to stakeholder expectations and values, including the sustainable performance of dams and reputational risk. The risk of impoverishment is linked to risk to healthy ecosystems and the ability of local people to manage ecosystem services from which they derive their livelihoods.[22]

An integrated sustainability assessment addresses many perceived failures to meet stakeholder expectations and to deliver the social and environmental performances of dams. Frequently, one underlying cause is the lack of money set out in operation-phase environmental management plans to pay for long-term measures or for integrating the project into the local economy and culture. In this context, sustainable performance means moving away from the high-risk strategy of sole reliance on overburdened state or municipal budget funding for long-term social and environment management measures. Instead, good practice means looking for sustainable financing to internalize these costs in bulk tariffs for water and energy services.

Investing in communication for sustainability is not only to support multi-stakeholder processes that inform the policy framework and project design parameters but also to inform the adaptive management of dams based on values important to the local community and people who use the river. Adaptive management is a way of thinking that must be translated transparently for all aspects of dam management, from bulk water supply agreements or power purchase agreements to arrangements for multistakeholder review of impact monitoring and statutory provisions to require operators to act on monitoring results.

Table 1.3 illustrates the differences in thinking and practice in terms of enabling conditions and adopting good practice to enhance sustainability in dam planning and management.

Communication as an Evolving Practice

Like engineering, environmental science, and economics, communication has different areas of practice—all applicable to project planning and management. The four areas of communication practice most relevant to contemporary dam planning and management are (1) development communication, (2) corporate communication, (3) advocacy communication, and (4) internal communication.[23]

Figure 1.3 illustrates the main aspects of each area of communication practice. Project managers should be broadly familiar with these aspects in order to better judge which can add value to their specific project when they assign priority to communication investments.[24] Each communication practice has a specific scope, approach, and set of tools.

Good communication practice on a dam project would start with a communication-based assessment that considers how each area of communication can help to advance project goals. The CBA determines for the practitioner what is essential (1) to understanding the risks that different stakeholder interests perceive to be important, (2) to building trust, and (3) to managing stakeholder relations successfully.

The Development Communication Focus

Figure 1.4 illustrates the four-phase development communication framework used in all World Bank operations, including dam projects. The four phases are communication-based assessment, strategy formulation, implementation, and evaluation.

The communication strategy for each stage of the infrastructure project cycle implies repeating the four-phase development communication cycle at the start of the operation phase, which will be discussed in more depth in chapter 2. Chapter 3 offers thumbnail sketches of tools and techniques for each phase.

Table 1.3. What Is the Difference between Thinking and Practice? Enhancing the Sustainability of Dam Projects and Balancing the Economic, Social, and Environmental Performances in Context

Typical past thinking or practice ⟶	Enabling or representing good practice
Change in thinking	
✂ Sustainability of dams is narrowly defined as environmental, financial, or structural integrity.	✓ Sustainability is widely defined as achieving balance in the economic, social, and environmental performance of dams. Stakeholders participate in negotiating or setting that balance.
✂ Dams are optimized as physical assets for the delivery of water and energy services.	✓ Dams are optimized for sustainable performance, overall development effectiveness, and equitable distribution of benefits and cost.
✂ Sustainability measures are financed over the long term by government budget allocations at municipal or higher levels.	✓ Sole reliance on government budgets is replaced by internalizing the cost of financing sustainability components in bulk supply tariffs.
✂ Adaptive management is viewed as a nice concept but essentially a luxury constrained by "real-world" needs such as contracts.	✓ Adaptive management is based on impact monitoring and multistakeholder review to inform regulatory oversight (e.g., on reservoir operating strategies and downstream releases).
✂ Supply and demand management decisions are made separately. Basin development is built around a hydropower development plan.	✓ Fundamental approach of demand-supply reconciliation is adopted, placing decisions about dams in the river basin (integrated water resource management) context.
Change in practice	
☹ Limited or narrow baseline studies are a low priority compared with "real action," such as site preparation works.	☺ Baseline studies are undertaken early, covering each sustainability dimension to provide performance improvement baselines, while consulting with people.
☹ Assessments of environmental, social, and economic impact/mitigation are compartmentalized or narrowly defined.	☺ Integrated sustainability assessment or linking existing sustainability assessments is undertaken, involving stakeholders in eliciting values and expectations.
☹ Stakeholders are engaged primarily through information dissemination, onetime surveys, and some consultation.	☺ Explicit mechanisms are adopted for multistakeholder involvement in the project governance structure empowered to report to regulators and public.
☹ Environmental and social management components are financed through government budget allocations.	☺ Benefit-sharing and revenue-sharing mechanisms are adopted to fund core budgets for all long-term sustainability aspects that people value.
☹ Compensation flow releases in water transfer projects are based on rule of thumb—for example, 10 percent of minimum flow.	☺ Environmental flow assessment and provision inform design (e.g., size of outlets, variable level intakes) and adaptive management.
☹ Impact is monitored, but there is no requirement for dam operators to act on results unless concerns can mobilize political pressure.	☺ Impact monitoring with stakeholder review is linked to regulatory systems that specify adaptive management of dams linked to the water quality-quantity status of river basins.

Figure 1.3. Four Areas of Communication Practice Relevant to Contemporary Dam Planning and Management

Development communication

- Structures design and supervision of communication programs around the four-phase development communication methodology (detailed in chapter 3)
- Explores and assesses the development context, builds a wider consensus among stakeholders, and assigns priority to communication activities
- Applies a communication-based assessment (CBA) to generate an overarching and multilayered project communication strategy to engage interested and affected parties in decisions on different aspects of a project and on implementation and operation
- Encompasses emergency or crisis communication otherwise minimized by investing in an up-front CBA to understand and address expectations.

Corporate communication

- Uses media and other methods to explain what a public or private enterprise is and does, not simply to boost image but also to serve stakeholders
- Consists of the procedures, rules, and "communication culture" the organization brings to dam planning and management
- Tells how the entity reflects government policy and how it will implement reforms on its dam projects
- Explains corporate ethics and how the organization will combat corruption, provide access to information, and pursue the sustainability agenda

Internal communication

- Plays an essential role in the internal governance for a project or the organization implementing a project
- Ensures the timely flow of accurate, relevant information so people can interact and do their jobs
- Achieves an internal consensus on goals and messages before going "outside"
- Is central to dam planning and management when multiple new organizations are created—as in water governance reforms with multiple new users
- Facilitates actual operation of projects in accordance with the assumptions made in their design

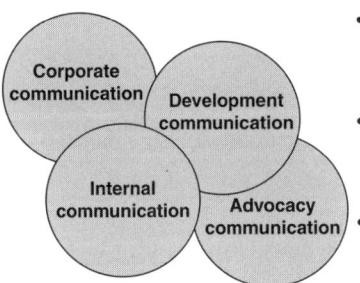

Advocacy communication

- Raises awareness and wins support with the public or influences relevant policy making
- Mobilizes a multistakeholder coalition to make the case for embedding anticorruption measures in dam projects or sustainability measures, such as environmental flow assessment and provision and benefit sharing.

What Is New or Significantly Different?

Good communication practice is universally recognized by communication practitioners as an integrated two-way process. This approach recognizes the strong views across a range of interests in dam projects, reconciling and managing stakeholder expectations.

Strengthening governance and sustainability in dam planning and management is more a change of emphasis than a radical departure:

- **More emphasis on the analytical components of communication.** Greater use of communication-based assessment tools, such as stakeholder analysis, public opinion research, and attitude surveys, to inform design of the project components.

Figure 1.4. The Four-Phase Development Communication Process: Activities and Outputs

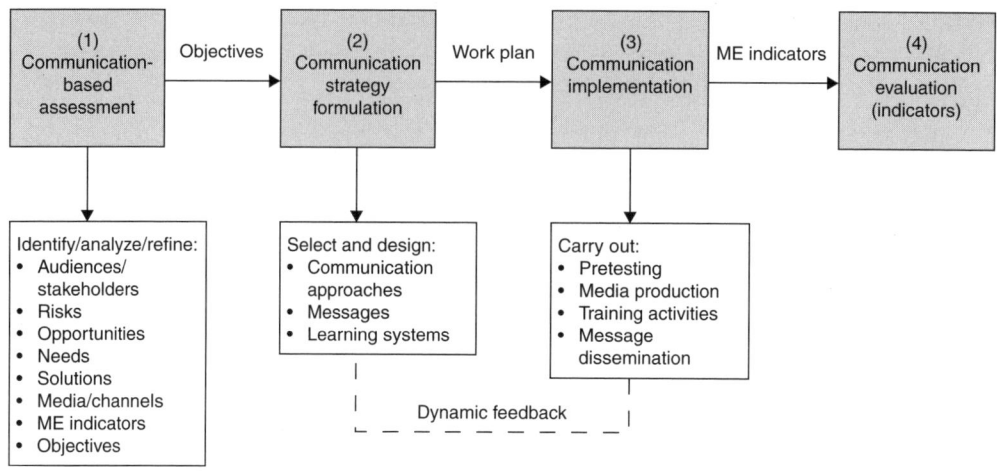

Source: Development Communication Division, World Bank.
Note: ME = monitoring and evaluation.

Figure 1.5. Analytical Support for Communication Needed through All Project Cycle Stages

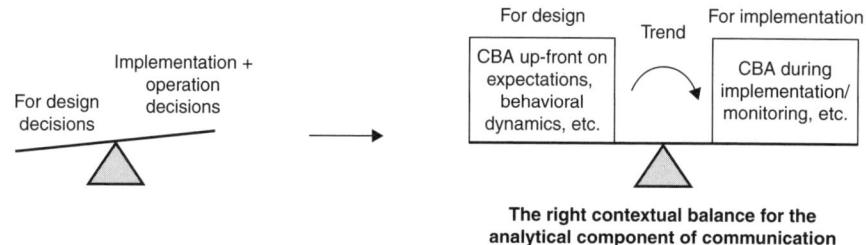

Source: Lawrence J. M. Haas.

- **More emphasis on advocacy of good practice.** Greater use of advocacy communication tools, which make the case for good practice, supporting the goals of anticorruption and sustainability.
- **More emphasis on communication in project implementation and operation.** Comprehensive communication support over all stages of the infrastructure project cycle, focusing not just on preparation but also on implementation and operation.

Why these greater emphases? The survey of Bank task managers for projects in Latin America noted in chapter 3 clearly indicates that after the Board approval stage of the World Bank project cycle the project-related communication investment often declines.

The communication-based assessment explores ways to extend communication support through the implementation and operation stages. These activities help the practitioner to monitor how multistakeholder partnerships are functioning, to evaluate and balance power relationships, and to detect and rectify potential problems in their early stages before adding significantly to project risk. Figure 1.5 illustrates this concept.

Ultimately achieving good governance and sustainability in dam planning and management entails adopting comprehensive communication strategies that use a wide range of tools, channels, and messages (table 1.4). Each of these activities targets specific needs in the wide range of stakeholders to support common project goals.

Table 1.4. How Each Communication Practice Helps to Advance Governance and Sustainability Improvements in Dams

Development communication	Internal communication	Corporate communication	Advocacy communication
1. Prepare an overarching project communication strategy using the four-phase development communication methodology.	1. Focus on how communication enhances internal project governance, coordination of actions, and multistakeholder partnerships.	1. Align communication strategies of public enterprises and private organizations involved in dam project with international and regional best practice and norms for corporate communication, including governance and anticorruption (GAC) and sustainability themes.	1. Each organization involved in dam planning and management advocates its delivery and support of beneficial change in the project in terms of government policies and the regulatory environment.
2. Benchmark the communication approach and tools used against the practice and share results with stakeholders.	2. Consider ways and means of improving communication (1) between stakeholders and multidisciplinary experts working on structural and nonstructural components of the project, (2) within project team and partner institutions, and (3) between the contractor and the construction work force.	2. Link risk management and sustainability objectives at the corporate and project levels to stakeholder values and expectations, respectively.	2. Advocacy builds support for incorporating corruption risk assessments and governance improvement plans, as well as new measures to advance dam project sustainability, such as environmental flow assessments and design aspects relevant to climate mitigation and adaptation.
3. Maximize coordination with environmental impact assessment (EIA) and the emergency preparedness plan (EPP) communication components to enable the public to participate in and to influence decisions that affect them.	3. Track how communication affects project outcomes and project perceptions by the public and stakeholders.	3. Adopt corporate communication strategies that explain policies to prevent and detect corruption, set out ethical behavior for working with stakeholders, and spell out disclosure and access to information policies.	3. Where feasible, project authorities offer to hold field trials for the most promising new methodologies that add value to the overall development effectiveness of the project and then transfer lessons to other dam projects in the basin, country, or region.
4. The four-phase development communication methodology is repeated to produce an operation stage communication strategy, drawing lessons from implementation stage communication.	4. Undertake a communication-based evaluation for risk management.		

Source: Lawrence J. M. Haas.

Table 1.5 illustrates the differences in thinking and practice in creating enabling conditions and adopting good practice communication in dam planning and management.

Table 1.5. What Is the Difference between Thinking and Practice? Modernizing Communication Practices in Dam Planning and Management

Typical past thinking or practice ⟶	Enabling or representing good practice
Change of thinking	
✂ Communication function for dam projects is mostly dissemination of information and public relations (PR).	✓ Communication goes beyond transmission of messages; it is a two-way dialogue to engage people in the development decisions affecting them.
✂ Unsupervised communication adds risk, has potential to stir up controversy, and opens "Pandora's Box."	✓ Communication is essential to advocating good practice, advancing understanding, and mitigating the risks each stakeholder perceives as important.
✂ Communication is important but not central to the delivery of water and energy services.	✓ Communication is essential to helping identify and adopt good practice across a range of issues and interests for sustainable infrastructure provision.
✂ Research and analytical components of development communication are nice if affordable, but they come on top of routine project management judgments.	✓ Analytical components are central to clarifying political and cultural dynamics and behaviors that hinder or advance reform, project goals, and functional multistakeholder partnerships.
✂ The "least-cost" strategy for communication investment in dam projects is to do the minimum needed for information dissemination and to respond to crises if and when they arise.	✓ A "least-cost" strategy for communication, and the project as a whole, is to measure and then manage expectations and empower multistakeholder partnerships that resolve and minimize conflict.
Change of practice	
☹ CBAs are not undertaken during project preparation to inform the project design.	☺ CBA tools connect decisions on dams to policy reforms at each stage of the project and help maintain multistakeholder partnerships.
☹ One project communication strategy is sufficient to cover all stages of the project cycle for dam projects, with minor updates.	☺ Communication strategies should be prepared for each project cycle stage, building on the previous stage as issues differ, expectations and risks evolve, people change, and actors change roles.
☹ Decisions on communication messages, strategy, and activities are the responsibility of project management.	☺ Communication activities should be benchmarked against relevant good practice and the results shared with stakeholders.
☹ Communication evaluations are rapid exercises generally not shared with stakeholders.	☺ Communication evaluations should be undertaken in close cooperation with stakeholders, and they should contribute to continual improvements in communication strategies to maintain a sustainable performance of dams in line with stakeholder expectations.

Capturing Synergy from Governance, Sustainability, and Communication

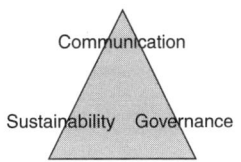

The notion that communication plays a key role in introducing governance reforms and providing sustainable infrastructure is not new. What is new is linking these mutually reinforcing elements in dam planning and management as an explicit strategy. This new strategy makes conceptual and practical sense; it helps to manage risks and expectations systematically.

Conceptually, the opportunity for three-way synergy stems from the following typical situations:

- Reforms, whether for anticorruption or water governance, are set within a participatory framework, requiring more open and inclusive decision processes.
- Governance and sustainability connect in many issues and interests. These involve the same interested and affected parties who inform decisions on competing interests or cooperate on shared interests.
- Synergy is achieved when the issues are tackled holistically, thereby achieving more for less. This approach avoids inefficiency and unnecessary confusion on the ground.

According to Transparency International's *Global Corruption Report 2008*, "Few things are more fundamental to sustainable development than ensuring that the management of the world's water resources is sustainable, equitable, efficient and free from significant governance failures, including corruption." This quote highlights the juxtaposition between corruption and sustainability in water management.[25] Communication supports anticorruption efforts, which in turn support sustainability by mitigating risks to the social and environmental performance of dam projects.

In real life, many of these issues are inseparable to people on the ground. For example, when the project authorities meet with host communities or river users to talk about one set of issues such as the environmental management plan, people will generally raise all of the issues that concern them. Controversial issues are not easily dealt with in isolation—confidence and trust are built by resolving the issues that affect the whole relationship.

The Berg Water Project, described in appendix A, is a clear illustration of synergy among issues important to stakeholder and project management. The multistakeholder environmental monitoring committee became the main platform for dialogue on resolving social sustainability concerns about the project, including the equitable sharing of benefits with the host community. In response to the concerns, the environmental monitoring committee (EMC) actually set up working groups to deal with aspects of project governance and sustainability. EMC members also agreed on a comprehensive communication protocol. This protocol guided member interaction, communication with respective constituencies, and engagement of the media.

Figure 1.6. Multiple Aspects of Governance Reform and Its Impact on the Policy and Planning Context of Dam Projects

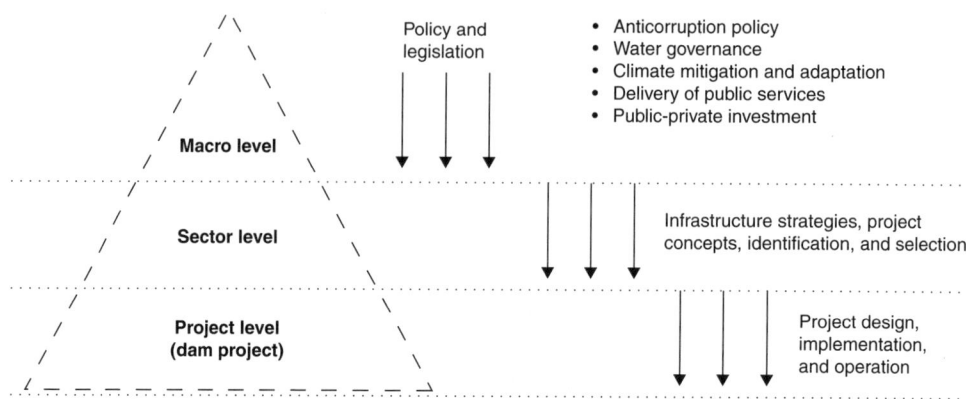

Source: Lawrence J. M. Haas.

Connecting the Project to the Governance Reform Agenda

Another practical aspect of three-way synergy is that it aligns different threads of governance reform evolving from infrastructure strategies to ensure that coherent approaches are followed to implement and manage what may be several dam projects throughout the basin or country.

In most countries, interconnected governance reforms under way at the macro and sector levels create a changing policy dynamic for dam planning and management, as illustrated conceptually in figure 1.6. Communication-based assessment diagnostics give task managers a better picture of the dynamic change and the need to link these aspects to project design and management decisions.

Communication analysis helps in understanding the important nontechnical or adaptive challenges that impede the success and sustainability of reform efforts. Conversely, it is important to understand which reforms can be tried at the project level to demonstrate the efficacy of reforms, thereby advancing innovative thinking and the grounding of sector-level and macro reform policies.

How Is Synergy Captured?

Figure 1.7 is a simplified illustration of how to capture this synergy. From the anticorruption perspective, the impetus for improving governance typically originates in the public sphere. A good understanding of the public sphere helps to use advocacy effectively in including anticorruption measures in specific dam projects and to decide the best approach operationally to introducing GAC measures. From the sustainability perspective, it is important to understand how the performance of the dam connects throughout the sector. Relevant questions might be: What role does hydropower play in the sustainable development of the power sector? Is it feasible to reduce the reservoir drawdown to improve environmental and social performance? How can one best meet greenhouse gas (GHG) or air quality emission reduction targets

Figure 1.7. Synergy Starts by Linking Communication-Based Assessment with Other Strategic and Technical Assessments

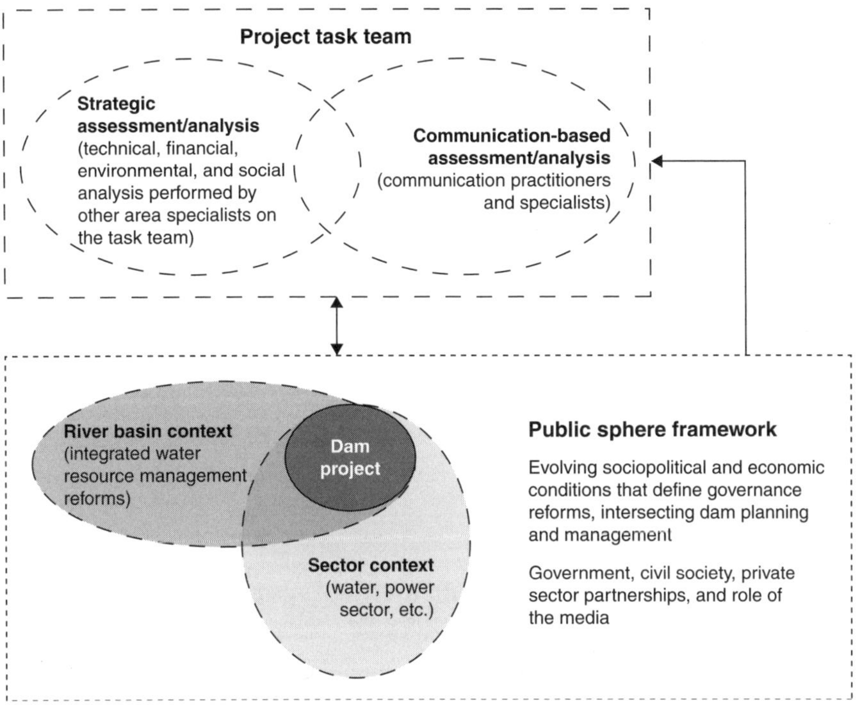

Source: Lawrence J. M. Haas.

while balancing energy needs or water security? From the river basin management perspective, it is important to understand how the operation of the reservoir complements policies to achieve sustainable management of water resources and how to connect and implement a range of policies for long-term management of wetlands systems in the basin.

Another aspect of figure 1.7 highlights the importance of linking the different practitioner assessments (e.g., sustainability assessments, corruption risk assessments, communication-based assessments) to identify and take advantage of synergy from an early stage in the project.

All of these assessments are an important contribution to the design of the project governance structure. They enable the appropriate mechanisms to capture synergy in the implementation and operation stages and to build it into the project through monitoring, multistakeholder review, and annual feedback meetings with the host community and river users.

Adding Value for Relevant Groups of Stakeholders

Like any investment in a dam project, investments in communication must add value. From different perspectives, communication and the governance-sustainability-communication synergy can add value along several dimensions.

Table 1.6. Added Value of Communication: Project Management and Stakeholder Perspectives

Project management perspective	Stakeholder perspective
Communication offers . . .	Communication offers . . .
• Improved project and portfolio quality and outcomes; alignment of dam decisions to evolving policy priorities	• Opportunity to better communicate expectations and have a voice in development decisions affecting stakeholders and their community
• Improved capacity to measure and manage expectations and risks	• Better service and multiple benefits flowing from reduced corruption and sustainable performance of dams
• Ability to achieve more for less; simplify responses to complexity by empowering more robust partnership approaches; improve trust	• Systematic attention to stakeholder-identified risks, which results in less frustration
• Ability to reach difficult decisions on dam projects; enhance political legitimacy; minimize costly delays	• Better integration of dam projects into the local economy and culture
• Improved public acceptance of infrastructure strategies and decisions on large dams	• Opportunity to view dam projects as major development interventions, which is a catalyst for local and regional development, leading to greater opportunity and empowerment, such as job creation
• Opportunity for task managers who use good practice to facilitate work with stakeholders and gather more resources, including support from Bank management	

Table 1.6 illustrates the perspectives of project management and stakeholders on the value added by communication.

If people feel they have been adequately consulted and involved in decision making, they will be more likely to support the resulting decisions. This theory has repeatedly received support in many fields and countries in the past, but it is relatively new to decision making for water resources and dam planning and management.

Based on Australian experience in water management, figure 1.8 shows the preferred level of public involvement in a partnership as opposed to community domination of decision making or top-down decision making by government or project authorities. Planning professionals also prefer partnership arrangements.[26]

The case studies in appendixes A and B illustrate the appetite of project-affected populations for adequate involvement in the decisions that affect them. Whether project authorities have the mandate to make management decisions is not the question; rather, it is the desire of stakeholders to ensure that their expectations to benefit are fully understood and realized to the maximum extent possible, recognizing there may be trade-offs with other stakeholders.

Reinforcing the Advocacy Roles of World Bank Task Teams

Advocacy is a vital aspect of Bank-supported projects in several respects.

First, as an international development institution that provides technical advice, guidance, and information to member countries, the Work Bank places

Figure 1.8. Community and Planners' Perceptions of Current Community Roles and Preferred Community Influence in Water Management Decisions

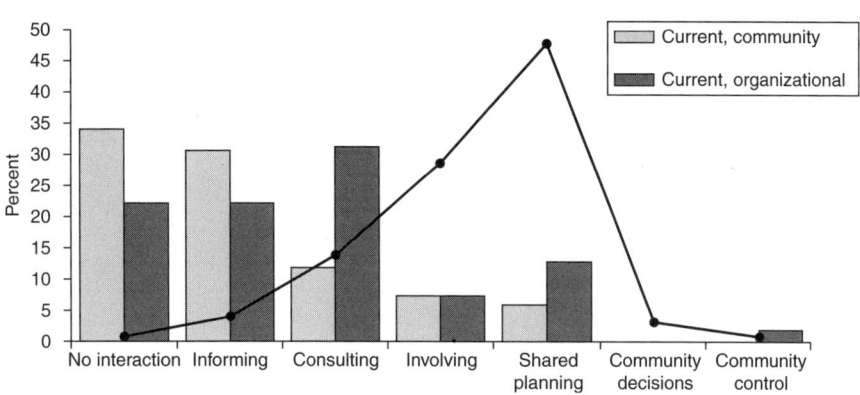

Source: Energy Sector Management Assistance Program (ESMAP), "Stakeholder Involvement in Options Assessment: Promoting Dialogue in Meeting Water and Energy Needs—A Sourcebook," World Bank, Washington, DC, July 30, 2003.

great stock in advancing good practice and continual improvement in many areas of development policy.

On the themes of sustainable infrastructure and anticorruption, the World Bank is at the forefront of developing good practice. The Bank's own internal reformulation of its Operational Directives into the three-tiered structure of (1) policy, (2) procedures, and (3) good practice reflects its priorities of generating and disseminating good practice.[27]

Second, it is in the interest of task managers to ensure the proactive promotion of good practice when working with client governments on Bank-supported dam projects, according to contextual needs. Advocacy recognizes the intertwining roles of various stakeholders—such as the government, international organizations, nongovernmental organizations, civil society groups, and the private sector—that are essential in the development and management of sustainable infrastructure, as well as anticorruption reforms that serve everyone's interests.

Advocacy starts with understanding the development context and continues with preparing key messages to influence perceptions and behavior in the early stages of project formulation and design. It is, however, also important during the implementation stages, which can span several years, because the operating arrangements of dam projects crucial to environmental and social performance are often decided closer to the project commissioning stage. Similarly for rehabilitation projects and operational improvements, various dimensions of advocacy can be considered.

Table 1.7 explains why steps to build anticorruption components into the design of dam projects are important from all stakeholder perspectives. It also shows why, to be convincing, an effective communication strategy should contain a common interest message to help form partnerships and coalitions.

Task managers should work closely with country clients to identify opportunities for building good practice measures in Bank-supported projects and to develop an effective advocacy strategy based on coalition approaches.

Table 1.7. Why Fighting Corruption in Hydropower Is in the Interest of All Stakeholders

Stakeholder group	Corrosive effects of corruption
Electricity consumers[a]	• Less affordable electricity • Less access to electrical services for poor and low-income people • Slower pace of expanding access to electricity in the country • Lower reliability of supply and levels of service
Communities affected adversely by hydropower development	• More high-impact or "bad projects" approved • Higher adverse livelihood impacts and impoverishment risks • Fewer funds for compensation, mitigation, and benefit sharing • Fewer commitments kept for the construction and operation stage mitigation necessary for sustainable management of hydropower asset
Electricity utilities	• Higher costs for bulk energy purchases or own supply • Higher borrowing and equity contribution costs for investments • Greater financial constraint on expanding electricity access and improved levels of service • Expensive to operate and maintain infrastructure and delays
Governments	• Higher power sector costs • Higher repayments for sovereign loans or guarantees • Setbacks to social policies and poverty alleviation • Lower rates of economic growth that depend on expanded or improved electricity services with a comparable reduction in job creation
Public hydropower developers and operators and independent power producers (IPPs)	• No level playing field for fair competition • Possible termination of licenses or approvals procured through bribes, resulting in collapse of project • Distortion in criteria for forming project consortium • Disqualification from office or criminal prosecutions
Contractors and E&M equipment suppliers	• Distorted and unfair competition for equipment suppliers • Higher and wasted tender expenses • Termination of contracts procured through bribes, resulting in the collapse of the contractor and loss • Criminal prosecutions, fines, blacklisting, and reputational risk
Financers: Export credit agencies, multilateral development banks, commercial banks, credit agencies, and insurers	• Higher reputational risks if corruption is proven on project support • Higher than necessary requests for borrowing • Additional costs and fraudulent claims • Financial loss

Source: Transparency International, extracted from the initial drafts of the Hydropower Section of the *Global Corruption Report 2008.* http://www.transparency.org/publications/gcr/gcr_2008.
Note: E&M = electrical and mechanical.
[a] On multipurpose projects, consumers include irrigators, urban water users, or any groups that would benefit from reducing corruption in water and energy service provision from multipurpose dams.

Notes

1. In the language of economics, derived demand is where demand for one good or service occurs as a result of demand for another.

2. A vast difference exists between agreeing on what constitutes beneficial change in communication practices and agreeing to commit the necessary financial and human resources, particularly because the task manager's aim is often to do more for less to meet budget constraints. Here it is necessary to overcome the resistance of some infrastructure practitioners to investing in modern communication. Their first instinct may be to see this investment as just another complication, or burden, rather than a derived demand that offers solutions to real-life complexity. It is clear that the challenges in proactively and simultaneously tackling corruption, enhancing sustainability, and improving communication around dam projects are intertwined, as are the many rewards.

3. See the World Bank's Web site on infrastructure financing, http://web.worldbank.org/WBSITE/EXTERNAL/EXTABOUTUS/ORGANIZATION/EXTINFNETWORK/0,,contentMDK:20535656~menuPK:1827891~pagePK:64159605~piPK:64157667~theSitePK:489890,00.html.

4. International Water Management Institute (IWMI, Colombo, Sri Lanka), "Water for Food, Water for Life: Insights" (paper prepared for Stockholm World Water Week, 2006).

5. These sectors are (1) realigning energy policies to reflect the urgency to move to low-carbon energy systems based on renewable energy resources, including hydropower, and meet emission reduction targets at the national and regional levels; and (2) advancing new financing models, such as public-private investment models for strategic infrastructure provision where the challenge is to find the proper balance between public and private sector risk allocation and responsibilities and public acceptance.

6. For example, to reduce global greenhouse gas emissions by replacing fossil generation with hydropower and providing the strategic flexibility to adapt to local manifestations of climate change by increasing water storage. In combination with demand management, additional water storage capacity relieves the pressure on existing water resource systems. It provides the climate with "headroom" to better adapt to climate variability and extremes.

7. According to George W. Annandale, president of Engineering and Hydrosystems Inc., the worldwide average annual rate of storage loss from reservoir sedimentation is on the order of 0.5–1 percent of total storage capacity. This amounts to having to replace some 300 large dams on an annual basis worldwide, at an estimated cost of $9 billion just to replace existing storage capacity (and not counting the cost to deal with environmental and social issues). Reservoir conservation and sediment management in reservoirs are an effective approach to maintaining existing storage capacity so that not as many new dams have to be constructed. It is also advisable to design new dams in a manner that will facilitate sediment management and long-term reservoir conservation. George W. Annandale, "Reservoir Conservation and Sediment Management," speech delivered to Water Week, World Bank, Washington, DC, April 2001.

8. The Bank has lent $20 billion for water-related projects in more than 100 countries. Projects support physical infrastructure as well as policy reform programs and cooperative approaches to building institutional capacity for the management of water resources, from the local to the transnational level. The Bank's Hydropower Business Plan (2008) foresees significant increases in hydropower development in many parts of the world, both for new facilities and for the rehabilitation and capacity upgrading of existing dam facilities.

9. There is also growing appreciation of the manifest linkages among water, environment, energy security, and climate change that require new channels for dialogue in developing infrastructure strategies that encompass both new dams and the rehabilitation and management of existing dams.

10. Many reforms, complex or simple, involve behavioral change. The need is the same whether introducing water governance reforms, IWRM reforms, or anticorruption measures. Modern tools and approaches are indispensable.

11. Transparency International, *Global Corruption Report 2008,* http://www.transparency. org/publications/gcr/gcr_2008.

12. But this finding specifically excludes corruption risks in service provision.

13. Although the causes and impacts of corruption vary from country to country, corruption in infrastructure is sometimes characterized as supply and demand–driven. Supply represents the behavior of the "payer" or "supplier" and the reasons for his or her behavior. Demand describes the behavior of the "taker" or the "demander" and the reasons for his or her behavior. Either side can initiate a corrupt act. The supply side thus encompasses the domestic private sector involved in infrastructure, as well as international actors. TI's Bribe Payers Index (BPI) ranks the propensity of private enterprises in particular countries to pay bribes. Instruments such as the OECD's Convention on Combating Bribery of Foreign Public Officials are aimed at the international actors. However, TI has criticized OECD's follow-up of the convention—apparently it has received only limited enforcement in at least two-thirds of OECD countries since it came into force in 1999. TI's Corruption Perceptions Index (CPI) documents a country's reputation for honest practice, and countries with an adverse rating appear to be high on demand-side corruption.

14. Some broader measures of reform that combat corruption are the following:
    - Ensuring that the basics of good governance are in place, such as the rule of law; participation, accountability, and definitions of basic legal rights, including access to defined public services standards; clear roles for the branches of government; and performance-based accountability.
    - Assessing service delivery performance as an important means of detecting corruption—and one that is far more effective than performance audits.
    - Empowering citizens by supporting bottom-up reforms.
    - Disseminating information and rights to information.

15. This paper was widely circulated in 2006 and 2007 to private sector, government, and civil society representatives such as Transparency International for comment prior to its presentation at a meeting of the World Bank's high-level Development Committee. Civil society and private sector feedback can be found at http://www.worldbank.org/gover nancefeedback. It is one basis for structuring the World Bank's current GAC initiatives.

16. Reports on the GAC consultations are collected at http://go.worldbank.org/ DZQVU82PY0.

17. In the absence of national anticorruption initiatives that quantify the level of corruption risk in the country, or any preexisting risk assessments in the sector, two indicator tools helpful at the project identification stage are Transparency International's Bribe Payers' Index, which ranks the propensity of private enterprises in particular countries to pay bribes, and TI's Corruption Perception Index, which documents a country's reputation on demand-side corruption.

18. Today, sustainable development is defined in legislation in most countries as it is in Vietnam: "Development that meets the needs of the present generation without compromising the ability of future generations to meet their own needs, on the basis of a close and harmonized combination of economic growth, assurance of

social advancement and environmental protection"(Article 1, revised Law on Environment Protection, 2005). This law, which came into effect in July 2006, defines sustainable development for all sectors of the economy, including the power sector and hydropower.

19. This is illustrated in the Berg Water Project case study in appendix A—for example, larger outlets, variable-level intakes, and statutory monitoring.

20. Even though governance is synonymous with anticorruption at the World Bank, under the wider definition of governance (cohesive policies, processes, and decision rights) other governance reforms are integral to sustainable hydropower development and to managing the broader water resource implications of dams, including water governance reform, institutional reforms, and tariff reform, to name a few. IWRM is highlighted in figure 1.3 because the Bank–Netherlands Water Partnership Program mission is to improve water security by promoting innovative approaches to IWRM, thereby contributing to poverty reduction.

21. There is an impetus behind EFAs. As illustrated in chapter 3, alternative frameworks, approaches, and methods for environmental flow assessment now inform decisions as varied as regulatory assessment and licensing procedures for hydropower, basin studies, design of outlet structures on dams, and the operating strategies of reservoirs that link the management policies of dams to the management strategies of downstream wetlands systems and water status indicators where statutory provisions require such.

22. In a 2000 report of the World Commission on Dams, *Dams and Development: A New Framework for Decision-Making* (http://dams.org), risks refer to the risks of all stakeholders in hydropower project development. The notion of risk extends beyond traditional engineering, financial, and economic risk (e.g., voluntary risks taken by governments, developers, dam owners/operators, and financiers) to the involuntary risk absorbed by people affected by a project and the environment as a public good. As stakeholders in a hydropower development, project-affected communities face risks— to their rights as well as their access to resources, livelihoods, and welfare—that must be minimized. Their risk exposures also need to be managed, much like risk management for voluntary risk takers. The notion of risk in this application is further advanced in the paper "Rights, Risks and Responsibilities—Scoping Report: An Approach to Implementing Stakeholder Participation," prepared in 2005 for the International Union for the Conservation of Nature (IUCN).

23. Various reports of the World Bank's Development Committee on communication support for operations define these four areas of communication practice as follows:
    - *Development communication.* Exploring and assessing situations, building a wider consensus among stakeholders, and applying communication media and methods for change in order to enhance project effectiveness, governance, and sustainability.
    - *Corporate communication.* Using media and other methods to communicate, among other things, what the organization does, its code of ethics, and how it deals with questions of transparency; how the organization builds trust and public confidence to engage in dialogue on projects it undertakes; and how the organization will respond to the relevant governance reforms. The procedures, rules, and "communication culture" of the agency or organization responsible for the policy planning initiative or project implementation fall in this category.
    - *Advocacy communication.* Communicating key issues effectively to raise awareness and to win support from the public or to influence relevant policy making.
    - *Internal communication.* Concerning oneself generically with communications within project teams, organizations, and the multi-entity governance structures involved in dam planning and management to ensure a timely and efficient exchange of information among the various teams, committees, units, departments, or staff.

24. This knowledge would also help to clarify what is expected from various project partners—in terms of communication activities and commitments—as part of the overall management of the project and of stakeholder expectations in an open and transparent manner.

25. Transparency International, *Global Corruption Report 2008* (Berlin: Transparency International, 2008).

26. A single line is presented in terms of what is wanted by both parties, because there was virtually no difference between what stakeholders wanted and what government planners wanted.

27. Good practices contain advice and guidance on policy implementation—for example, the history of the issue, the sectoral context, the analytical framework, and best practice examples.

How to Adopt Good Practice

Five Key Messages on "How To"

1. The **communication-based assessment (CBA)** is the first opportunity to identify how communication might improve aspects of governance and sustainable performance and mitigate risks and to identify the advocacy actions needed to establish a consensus and the political will to integrate these aspects in the project design.

2. **Benchmarking must be part of the preparation** of the communication strategy and major communication products. Performance and process criteria can be used to benchmark against accepted good practice. Generic criteria are the following: (1) Have stakeholder expectations been measured and then managed? (2) Have the relevant stakeholder voices been empowered? (3) Has trust been built for functional partnerships? (4) Has value been added for different groups of stakeholders? (5) Have the tools for risk reduction—all forms of risk—been adequately advocated?

3. **Communication strategies should be conceived as public documents,** shared with or owned by stakeholders, who must be given an opportunity to provide inputs and comments. Typically, each major project activity and partnership group would have an overarching and detailed communication strategy. One cannot assume that the communication strategy on the last project will be the benchmark.

4. **Communication strategies must be elaborated for each stage** of the infrastructure project cycle: planning, implementation, and operation. In most cases, updated communication strategies are needed for several reasons: (1) issues and risks change between project stages; (2) stakeholders' roles change; and (3) typically there is a long period of time between stages, because implementation can take several years. Similarly, the Development Assistance Committee (DAC) of the Organisation for Economic Co-operation and Development (OECD) advocates that communication strategies cover and be updated at each stage of the World Bank project cycle (a lending cycle).

5. **Communication improvements must deliver value for money and value for stakeholders.** Investments in project communication should be balanced against other project investments. Multistakeholder processes are the best way in which to inform judgments on the relative value that investments in communication add and how essential they are to all forms of risk management and meeting stakeholder expectations.

Starting with a Communication-Based Assessment

World Bank task managers have varying degrees of experience with modern communication theory and practice. How communication is handled within the Bank's own growing portfolio of dam projects is shaped by a variety of contextual factors on a case-by-case basis:

- Actions that the legislation and policies of client countries require in terms of information access, notification, and direct involvement of interested and affected parties in decisions about the design, implementation, and management of dams
- Procedures, degree of commitment to public participation, and the "communication culture" of the public or private enterprise responsible for dam planning, implementation, and operation
- Capacity of the media and civil society at the national and project levels and their experience in reporting on governance reforms and development issues associated with dam projects
- Attitudes, procedures, and practices of institutional partners related to public communication and the partners' behavior in partnerships
- Collective pressure of interested and affected parties for two-way communication and advocacy to enhance participation, transparency, and accountability
- Past failures and deficiencies in project implementation directly attributable to poor communications.

Often a particular project or context presents special communication needs. For example, a past controversy or historical grievances may define local attitudes toward the project—in postconflict settings often a postwar legacy of intercommunal strife or dislocation must be overcome. Good practice involves accounting systematically for all of these issues in the up-front communication-based assessment. Box 2.1 describes what the traditional CBA typically seeks to achieve in dam projects.

Box 2.1. What Is the Traditional Communication-Based Assessment (CBA)?

A CBA is a family of techniques to assess a dam project's local capacities, to gain insight into sociopolitical concerns and roadblocks, to understand perceptions of risk, to probe fears and expectations, and to determine the knowledge level and perceptions of the media on the project and related issues.

The purpose of the CBA is also to learn what partners and nongovernmental organizations (NGOs) are doing in the same area and verify the availability and skills levels of national research and communication agencies.

Finally, CBAs rely on qualitative and quantitative data to better understand the context of a dam project and to determine stakeholders' perceptions, opinions, and beliefs.

Table 2.1 is a simple checklist to illustrate the range of communication issues that may factor into different settings. Task managers can quickly assess whether the concerns are relevant to their project—the more checkmarks the greater the returns from investing in a proper communication-based assessment.

Investing in a CBA is an investment in risk management. Contemporary risk management philosophy takes into account risks to stakeholder interests, whether it is risk of corruption or failure to meet stakeholders' expectations in sustainable performance.

Task managers begin by demonstrating a disposition to listening, respect for the stakeholders, and an interest in genuine engagement as development partners.

Table 2.1. A Communication-Based Assessment: A Quick Checklist

Does the dam project or initiative . . .	Do institutional partners . . .
✓ have one or more controversial aspects that need to be reconciled with stakeholders? ✓ require innovation and new approaches to tackling governance and sustainability issues? ✓ need to bring people and ideas together across many different sectors, interests, or political jurisdictions? ✓ involve significant trade-offs between economic, social, and environmental performance, balancing competing rights?	✓ have corporate policies for prevention and detection of fraud and anticorruption, including codes of conduct? ✓ have a track record on governance improvement and sustainable projects? ✓ benchmark their communication, governance, and sustainability policies against accepted good practice? Routinely? Never? ✓ share the attitudes and culture of their partners important to communication?
Does the policy environment . . .	**How communities view communication**
✓ require linking dam decisions to policy reform processes (e.g., water and electricity governance, anticorruption reforms)? ✓ have requirements for public participation processes and interested and affected party involvement in the governance structures of the dam project? ✓ require project authorities to ensure access to information on all project aspects? ✓ have governance and anticorruption reforms under way to link to or draw lessons from? How does the country rate in terms of corruption risk indicators?	✓ Do local host communities want to be involved in decisions that affect them? Is there significant resettlement? ✓ Do civil society and interested and affected parties want to be involved in compliance and monitoring? ✓ Do existing water users in the river basin want to be informed and consulted on design and operation decisions? ✓ Do key project design and operation decisions concern water management institutions and water user associations in the basin? How do they confer and reach consensus?
Advocacy is needed . . .	**Are there special communication needs . . .**
✓ to include aspects of anticorruption in the design of the project. ✓ to improve public participation and stakeholder information access. ✓ to introduce new sustainability measures such as environmental flow assessment and benefit sharing with local communities. ✓ to adopt any aspect of good practice the World Bank promotes as a catalyst or model to advance for other dams in the basin. ✓ to advance any sort of advocacy products available and seek new products.	✓ to form and support the effective workings of multistakeholder partnerships? ✓ to facilitate conflict resolution on specific implementation or operation issues? ✓ to enable the local and regional media to function effectively? ✓ to work in multiple languages and culturally sensitive ways? ✓ to build communication capacity with different water user groups and organizations, informing decisions on dam building, design, and long-term management?

One practical aspect is to link the different needs assessments conducted on dam projects. For example, the CBA must be linked to the social assessment conducted in the framework safeguard studies for the project, as well as to the corruption risk assessment (CRA) and others relating to the project's sustainable performance risk. These are some of the elements that encompass the broad definition of stakeholder-defined risk.

Table 2.2 is a compilation of a generic list of "dos and don'ts" in assessing the project communication needs not only in project preparation but also along different stages of the project cycle.

Applying Criteria to Benchmarking for Good Practice

No formal standards for communication exist as yet for Bank-supported dam projects or, more generally, for dam planning and management. Until standards are available, benchmarking against accepted good practice is a practical step that task managers can take to systematically identify and implement better or the best communication practices.

Benchmarking is a widely accepted management tool to improve the performance of organizations, regardless of the field.[1] It is a tangible commitment to the continual improvement of practice. On dam projects, it applies not only to communication but also to practice in governance, anticorruption, and sustainable performance.[2]

There is considerable methodological experience on which to draw. Many academic and practitioner networks provide information on and tools to benchmark performance indicators and business processes against a variety of aspects such as social accountability, corporate governance, and anticorruption measures.[3]

How Is Benchmarking Communication Relevant?

Benchmarking, which can be formal or informal, offers a structure for considering improvements in communication practices in a collaborative way. Benchmarking communication practices on dam projects specifically helps task managers do the following:

- Obtain value-added feedback from stakeholders, triggered by giving people the opportunity to see and to learn firsthand how other dam projects and stakeholder groups respond to communication challenges
- Better judge the value of the incremental investments that communication offers, thereby reconciling investments in communication with other calls on project budgets
- Firm up the business case for improvements in communication and, by extension, support improvements in governance and sustainability.

What Is the Best Approach?

On Bank-supported projects, the communication specialist would routinely add a benchmarking section to the main communication management

Table 2.2. Some Dos and Don'ts in Starting a Communication-Based Assessment (CBA)

Don't do this	Do this
✂ Don't assume at the planning stage that all stakeholders understand how the dam project will affect them.	✓ Do undertake a communication-based assessment, giving due emphasis to governance and sustainability improvement opportunities along the project cycle.
✂ Don't assume it is "least cost" or a good risk avoidance strategy to keep a "low profile" and deal with problems if and when they arise, using crisis communication.	✓ Do ensure that the terms of reference for the CBA reflect governance and sustainability improvements, and do involve the task team appropriately in discussion of the CBA scope.
✂ Don't forget to assess the scope for advocacy activities to influence thinking and decisions positively to drive improvements in governance and sustainability.	✓ Do ensure that the CBA covers the requirements for public participation, routine information flow, and the communication needs of partnerships for project implementation and operation.
✂ Don't forget about advocacy for other good practice elements promoted by the World Bank and implicit in the government policy framework but not yet in practice.	✓ Do cover all aspects of advocacy, internal communication, and corporate communication—also with institutional partners.
✂ Don't forget communication between the contractor and contract workforce. This workforce is often the main contact with the local community, and this relationship involves many social sustainability issues.	✓ Do ask communication specialists to benchmark the CBA and the needs it identifies against emerging national and regional good practice and with other dam projects in the Bank's portfolio.
✂ Don't forget about support for the communication capacity of major stakeholder interests and partnerships, especially empowering local action to improve the overall effectiveness of project development.	✓ Do discuss the draft CBA with multistakeholder groups in the project governance structure with an open mind to solicit their feedback and suggestions for improvement.
	✓ Do use the CBA as a first building block to create a culture of communication on the project and to build stakeholder confidence and good faith in the implementing agency.

reports—for example, initially it would be part of the CBA, then the draft communication strategy, and eventually the evaluation phase report. The benchmarking framework would be discussed openly with stakeholders at an early stage.

The following five generic, crosscutting criteria form a simple framework to benchmark communication practices for improvements in governance and sustainability:

1. Have stakeholder expectations been measured and then managed?
2. Have the *relevant* stakeholder voices been empowered?
3. Has trust been built for functional partnerships?
4. Has value been added for different groups of stakeholders?
5. Have the tools for risk reduction and benefit enhancement been adequately advocated?

Good practices that respond to each of these criteria can be identified.

How Are the Generic Criteria Applied?

These or any other criteria can be systematically incorporated into the four-phase development communication methodology. As illustrated in figure 2.1, the process used to benchmark the communication strategy takes into account the relevant good practice, alongside the standard questions and CBA analysis used to formulate the project communication strategy.[4] The dimensions of improvements in governance and sustainability receive explicit attention. The complexity depends on the situation.

Two-way communication with stakeholders will reveal their perspectives on good practice that adds value. The communication strategy can then be modified to reflect stakeholders' views.

With allocation of time and resources, task managers can establish a small group, facilitated by the communication specialist, to answer the key question, "What needs to be benchmarked?" This group then helps to identify which aspects of a project should be emphasized in benchmarking reports that all stakeholder representatives would review collectively.[5]

A communication subgroup could undertake this task, or a "benchmarking team" composed of interested members drawn from a multistakeholder body already attached to the project. The team could be an existing stakeholder consultative committee or an environmental monitoring committee, such as the environmental monitoring committee (EMC) established for the Berg Water Project and now required for all water projects in South Africa, as described in appendix A.

On dam projects it is important, at a minimum, to benchmark the following:

- Communication-based assessment—scope, detail, and orientation
- Communication strategy and action plan—all relevant aspects
- Indicators of improvement in performance over time—related to the five generic criteria offered in this handbook and others deemed appropriate
- The approach to ensure independent evaluation of the communication function and self-evaluation as appropriate
- Engagement with the international, national, and local media
- Communication mechanisms at all levels within major partner institutions and interested and affected partners at the grassroots level.

Another aspect of good practice is how project governance structures and institutional arrangements help or hinder flexibility. This aspect includes practices to create rapid responses to communication challenges, to ensure messages are not overly complex, and to set out clear lines of responsibility and accountability for communication messages and context.

The establishment of communication protocols should be considered. These protocols can ensure clarity on the communication roles of the actors in public participation processes (PPPs) and of individual institutional partners in key project partnerships. Communication protocols can clarify (1) issues such as responsibility for dealing with the media and on what issues and (2) the

Figure 2.1. Benchmarking Communication Strategies within the Four-Phase Development Communication Process

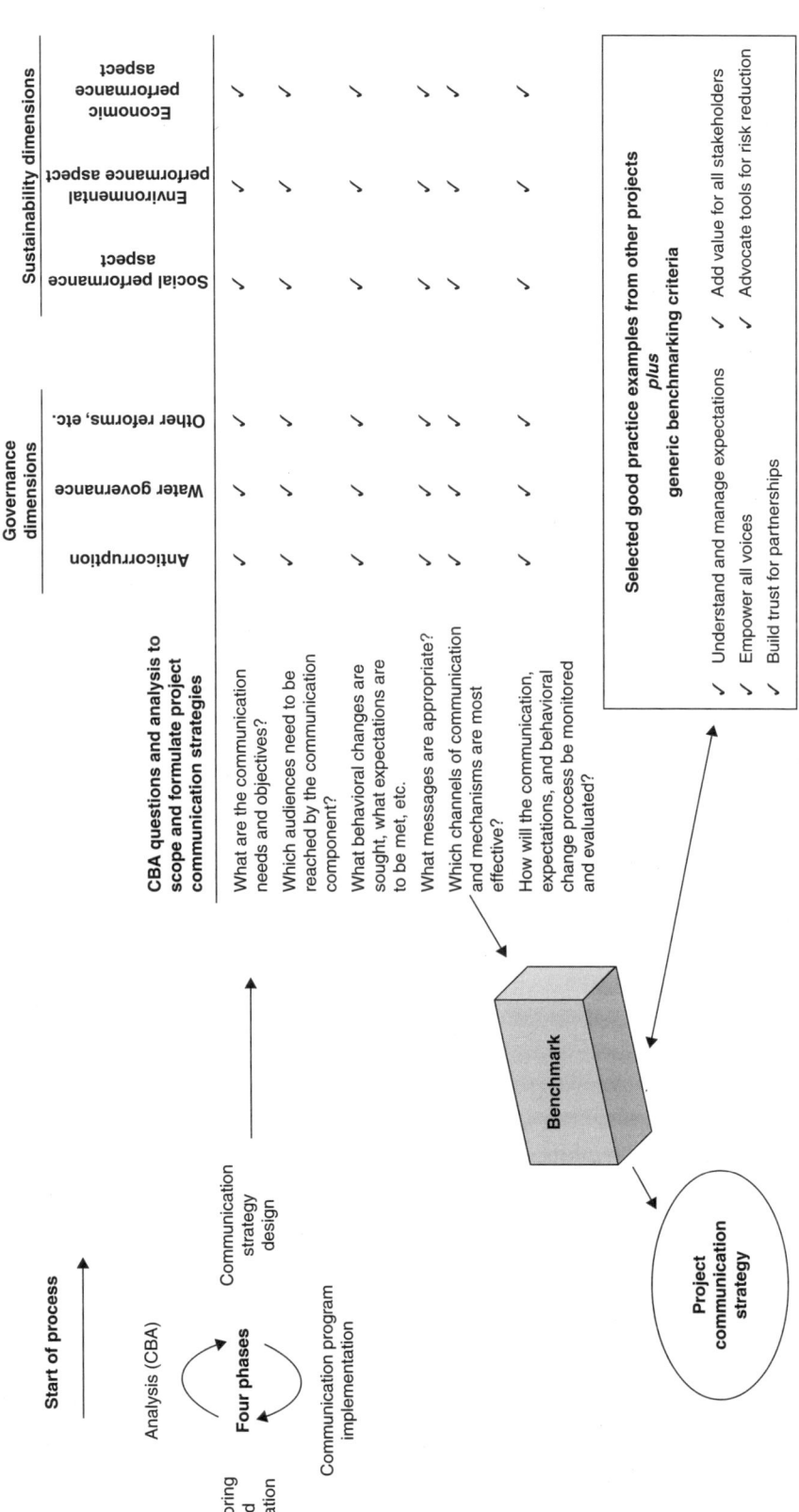

Source: Lawrence J. M. Haas.

responsibilities of individual stakeholders to faithfully represent the views of the constituencies represented.

The following sections describe how each of the five generic criteria is relevant to benchmarking good practice communication consistently. Each criterion is placed in the context of the Berg Water Project (BWP) and the Lesotho Highlands Water Project (LHWP). The overall section concludes with a list in table 2.3 of dos and don'ts on benchmarking good practice communication for dam projects.

Criterion 1: Measuring and Then Managing Stakeholder Expectations

Failure to adequately measure and then to manage stakeholder expectations is a common misstep on projects. Managing expectations is not a cynical act of manipulation. Rather, it represents understanding what all stakeholders expect of a project and what they fear to lose, and then managing the risks people perceive as important. It also means not overpromising to deliver benefits and then failing to deliver.

Controversy is fueled when the expectations of one or more stakeholders is not recognized from the start, or is not accommodated in the project design, or is simply not viewed as "important" in the scheme of things as the project proceeds. For partnerships to work, people must feel their expectations can be met or at least understood. This factor applies to all levels of partnership on dam projects and to all topics, ranging from collaboration on preventing and detecting corruption or revenue-sharing arrangements to enhancement of social sustainability during dam operation.

Dialogue mechanisms that enable stakeholders to better understand each other's views and expectations must be present. These mechanisms facilitate "good faith" negotiations when they are needed. Project authorities must listen and be proactive.

Communication-based assessment tools gather information about stakeholders' knowledge, perceptions, and attitudes. These tools help practitioners understand expectations. And they equip task managers with "radar" to identify potential problems early and manage them better.

Box 2.2 highlights how important measuring and managing expectations was to community acceptance and perception of outcomes on the BWP and LHWP.[6]

Criterion 2: Empowering Relevant Voices

This criterion reflects thinking that few people contest. Success in enhancing governance and sustainability hinges on effective coalitions of the government, private sector, civil society actors, and people affected by projects. These coalitions not only advocate appropriate measures but also define and implement them contextually. It is widely accepted that giving people a voice in the development decisions that affect them is just and equitable. It also makes practical sense on several levels:

Box 2.2. The Multiple Expectations Associated with Dam Projects

Berg Water Project. Within the host community in Franschhoek, South Africa, there were vastly divergent expectations of the project. Affluent and influential groups in the community wanted to minimize the adverse impacts on heritage, tourism, and the local environment. The poor and once economically marginalized groups in the community saw the effort as a public works project that would provide jobs during construction and operation. Accounting for all stakeholder expectations was fundamental to gaining public acceptance and to finalizing the project's components and communication support. Among the innovative measures emerging from the dialogue with stakeholders was the "Franschhoek First" policy of maximizing local employment and the economic spin-off during construction. This spin-off included a housing trust for the longer term, capitalized with money from the sale of construction phase housing.

Lesotho Highlands Water Project. Project proponents in both Lesotho and South Africa had expectations about the distribution of the benefits from binational cooperation. These expectations were captured in the original project treaty negotiated in 1988. In 1998 the treaty's terms were amended to reflect mutual agreement on the changed political conditions. Expectations in the traditional Highland communities had evolved over time. Annual stakeholder conferences are now held with local communities and project management. Prior to these conferences, all stakeholders rate project performance on issues related to social and environmental sustainability (Balanced Score Card) to measure whether individual expectations are being met. Management must respond.

- Two-way channels of communication are essential to providing stakeholders with genuine opportunities (1) to contribute expert and local knowledge to improving the development performance of dam projects, and (2) to participate in a meaningful way in multistakeholder partnerships.
- These partnerships raise many questions about the optimal development and management of dams. To some extent, trade-offs on different aspects of sustainable performance are negotiated, if not resolved. Decision authorities at all levels are better informed to make trade-offs when voices are empowered.
- Most governance and anticorruption measures at the sector and project level require active engagement with legitimate NGOs and civil society organizations (CSOs) and feedback from the consumers of services.
- Inclusive approaches and partnerships require legitimate representation (e.g., spokespersons nominated by each interest group) and enlistment of trusted public opinion leaders, if necessary. Fundamentally, a set of rules and a framework are needed to empower the relevant voices in order to nurture and develop the appropriate communication capacity.

Box 2.3. Using Rules and Dialogue in the Project Governance Structure to Empower Relevant Voices

Berg Water Project. A multistakeholder environment monitoring committee was the primary step in formalizing the voice of interested and affected parties in the BWP governance arrangement. The environmental monitoring committee (EMC) had an independent chairperson. Its 20-person membership consisted of representatives of the project authority, local and provincial government bodies, and various interested and affected party groups that included previously disadvantaged communities, ratepayers, and commercial interests in Franschhoek's tourist trade, as well as downstream water user interests (such as irrigation boards in the process of converting to multiuser associations), farmers groups, industries, and conservation and recreational groups. Each constituency elected representatives to the EMC, which then formed seven subgroups for specific functions. One subgroup was responsible for developing a communication protocol that all EMC members endorsed (see appendix A).

Lesotho Highlands Water Project. The governance structure of the LHWP included the Lesotho Highland Commission (LHC), which provided political oversight of the Lesotho Highlands Development Authority (LHDA). Five members of the commission were appointed to speak for the interests of the respective governments. Although the LHDA has a hierarchy of community representatives and processes for ongoing dialogue with local communities, the Office of the Lesotho National Ombudsman also backstops these processes and enforces the "rules of good faith engagement." The ombudsman is empowered to intervene and arbitrate disputes between local communities and the LHDA and does so often (see appendix B).

Box 2.3 describes the material steps used by the BWP and LHWP to empower relevant voices and how this empowerment was accommodated in project governance structures to benchmark progress.

Criterion 3: Building Trust for Functional Partnerships

Trust is one of the most essential ingredients of successful partnerships. Partnerships are both a catalyst and vehicle for the cooperation needed to optimize the development performance of dam projects. From a communication standpoint, neither general information nor targeted messages are absorbed if stakeholders do not trust the source or their partners.[7]

Many factors go into creating trust among stakeholders in dam projects: transparency, commitment to participation, culturally appropriate communication and feedback, and written commitments that can be verified and upheld. When projects have a conflicted history, steps must be taken to address legacy issues.

Experience shows that trust does not come automatically; it has to be earned. Trust is dynamic not static, and, once lost, it can be very difficult to recover.

In multistakeholder partnerships, trust is key to creating the space for agreement on sharing costs and benefits (within the policy framework and regulation) and on pursuing adaptive management of dams. Trust between riparian states is centrally important for mutually beneficial cooperation on international rivers.

Task managers also need to ascertain the level of public trust in government agencies, as well as in the public or private sector entities that implement dam projects and the regulators who may step in when unforeseen events occur.

Establishing a policy of clear two-way communication from the start is essential to creating, building, and maintaining trust—not only stakeholders' trust in authorities to act in "good faith" but also their trust in the mechanisms to reconcile stakeholder interests.

Criterion 4: Adding Value for Different Groups of Stakeholders

This criterion focuses on two aspects of adding stakeholder value through communication. First, any investment in communication must yield value for the money spent—that is, steps must be taken to maximize the impact of any communication activities. Second, communication must underpin inclusive decision making on dam design, delivery, and performance to ensure that they add value for different groups of stakeholders, not just some. Sharing benefits must be equitable and perceived as such.

Value is also linked to the notion of doing more for less—that is, delivering services through sustainable infrastructure while minimizing costs. This approach includes guarding against corruption-related cost escalations and understanding the range of values that governments and stakeholders attach to different sustainability outcomes.[8]

More fundamentally, value is about viewing dams as a development intervention, as a catalyst for poverty reduction and local and regional development, and not just as a way to optimize physical assets for water and energy service delivery. Communication strategies should be aimed squarely at delivering on stakeholder expectations.

In most developing countries, attention to value added is critical for host communities and affected river users, who typically live in remote rural settings. This is value as they perceive and define it—that is, an addition to the value added for the primary beneficiaries of water and energy services who typically reside in towns and urban centers.

Box 2.4 describes how communication investments in the BWP and LHWP were geared toward adding stakeholder value at all stages of the project cycle, both in extending the definition of value and in recognizing all stakeholders.

Box 2.4. Using Communication to Create Stakeholder Value at All Project Stages

Berg Water Project. Communication was central to involving the basin communities in BWP decisions at all stages. Media coverage of public participation in the strategic demand-supply reconciliation studies and project-level environmental impact assessment provided political legitimacy for the government decision to advance the project—conditional on a 10 percent reduction in water demand. Public communication was important to explaining the water tariff increase in Cape Town when the Berg Water Capital Charge (BWCC) was introduced. Specifically, it explained the link between tariff measures for water demand management and the equity in the application of the tariff increases. Corporate sustainability reporting by the Trans-Caledon Tunnel Authority, the BWP implementing agency, was central to raising concessionary financing. Favorable credit ratings brought both lower water tariffs and value for all stakeholders.

Lesotho Highlands Water Project. Communication was a very small component of the project budget in Phase IA of the LHWP, starting in 1991 with construction of the Katse and Muela Dams and power and water transfer facilities. By 2008 communication and public stakeholder consultation activities for Phase II accounted for more than a third of the feasibility stage budget. Lessons from Phases IA and IB revealed that many factors hinged on effective communication, including the political legitimacy of decisions, community acceptance of the project and understanding of how it stood to gain, external perceptions of the project, and assurances to investors that measures for governance and anticorruption were in place to prevent, detect, and avoid a replay of the well-publicized corruption problems in Phase IA.

Criterion 5: Advocating Forms of Risk Reduction

Communication must support the identification and management of different forms of risk, with risk defined in an encompassing way, not narrowly. Communication can support a more holistic approach to risk assessment and management, starting with a dialogue platform from which stakeholders can inform decisions on risks, and communication activities to enable stakeholders to help manage risks.

In practice, communication means ensuring that the risk exposures important to stakeholder constituencies are identified, properly assessed, and reasonably addressed from the outset of project preparation. Accurate identification of stakeholder expectations leads to greater accountability for risk management and better use of partnerships to manage the variety of risks.

Box 2.5. Tools for All Forms of Risk Reduction

Berg Water Project. The multistakeholder environment monitoring committee was a de facto "integrated risk assessment" platform for the BWP. Members transparently identified the risks their constituencies perceived as the most important, informed decisions on how to manage them, and agreed on stakeholder roles. As the implementing and operating agency, the Trans-Caledon Tunnel Authority (TCTA) also formally linked its corporate governance, risk management, and compliance objectives to the project funding and implementation philosophy: "Respond to all types of risk in all parts of the organization and business." The TCTA routinely benchmarked its corporate risk management practices, including anticorruption measures, against generally accepted risk management "best practice." One key benchmark used by the TCTA during the BWP implementation phase was the 2002 report by the King Committee on Corporate Governance (the King II Report).[10]

Lesotho Highlands Water Project. In Phase IB of the project, the LHDA introduced a program of environment flow assessment and provision into the management of releases from the Katse Dam. This approach recognized the risks to the livelihoods of the traditional river users downstream. The program incorporated compensation, which is to be reassessed after 10 years. The well-publicized prosecutions and disbarment of international firms during Phase IA activities were another crucial part of the overall risk management approach.

Important on this list are the risks that only local communities face. As local partners and risk absorbers, their voices are critical to project sustainability, taking into account "recognition of rights" and "assessment of rights at risk." It is a matter of focusing not just on the voluntary risk takers such as contractors, owners, and financing agencies but also on communities.[9]

Communication activities must facilitate participation in risk assessments across a range of topics, from livelihood risks that relate to social sustainability, to environmental risks, to risk assessment for emergency preparedness programs. Meanwhile, those in charge of these activities must develop strategies to make risks transparent and understood.

Risk to what is often called an organization's primary intangible asset— reputation—is also related to the risk of corruption. In major water projects, reputation is a vital concern for all parties and partners: developers, contractors, and financing institutions, as well as for the CSOs and NGOs that participate in dam projects.

Box 2.5 describes how important expanded thinking on risk management was to the project outcomes of the BWP and LHWP.

Table 2.3. Some Dos and Don'ts on Benchmarking Good Practice Communication on Dam Projects

Don't do this	Do this
✄ Don't assume that benchmarking is another 1990s management fad well past its "sell by" date.	✓ Do use benchmarking at the outset of the communication-based assessment and draft communication strategy.
✄ Don't assume that benchmarking applies only to business processes.	✓ Do use the five-criteria framework offered in this handbook, supplemented by any other criteria proposed by the communication specialist—be comprehensive.
✄ Don't forget to look for communication practices in the country and region that may be relevant to benchmarking.	
✄ Don't forget to check whether other good practice topics that the World Bank promotes in its own dam portfolio are relevant to the current project.	✓ Do involve the task team in the benchmarking exercise from the outset; present benchmarking as a stimulant for innovative thinking, not as a compliance tool.
✄ Don't forget about benchmarking advocacy communication steps, such as incorporating governance and anticorruption measures in dam projects.	✓ Do provide an adequate opportunity to discuss the proposed benchmarking framework at a meeting of stakeholder representatives at an early stage. Be open to stakeholders' suggestions.
✄ Don't forget to benchmark the communication support required for governance and sustainability tools.	✓ Do put the results of benchmarking to good use—for example, to refine a needs assessment or communication strategy.
✄ Don't forget about involving partners in benchmarking processes from the start and sharing benchmarking results more widely with stakeholders.	✓ Do consider benchmarking for corruption risk assessments and governance improvement plans and activities.
✄ Don't restrict benchmarking to communication alone; benchmarking should be applied to aspects of governance and sustainability improvement.	✓ Do consider benchmarking for new sustainability themes, especially local benefit sharing and environmental flow assessment and provision.
✄ Don't forget that benchmarking is a continual process and a commitment to continual improvement. It should be updated formally on a periodic basis.	

Formulating Comprehensive Communication Strategies

A project's communication strategy is the central document in managing the communication function on dam projects. On Bank-supported projects, the strategy is produced in the second stage of the four-phase development communication process depicted here.

Analysis (CBA)

Monitoring and evaluation

Four phases

Communication strategy design

Communication program implementation

Because dam projects incorporate many separate activities that concern different groups of stakeholders, a concise overarching communication strategy is typically prepared. This strategy links to the main project objectives and major expectations of stakeholders. It also provides key messages about project implementation.

Supporting this overarching strategy is a hierarchy or series of detailed communication plans for each significant project activity. Figure 2.2 illustrates

Figure 2.2. Plans and Subplans of Dam Projects That Need a Communication Strategy

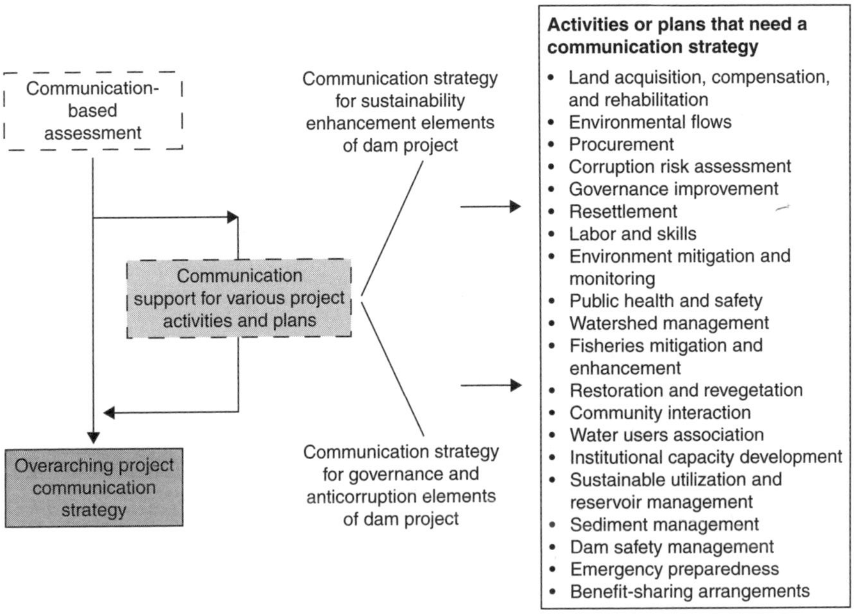

Source: Lawrence J. M. Haas.

the range and types of activities that require supporting communication strategies and components on dam projects.

For the governance dimension, a corruption risk assessment engages stakeholders in a collaborative way. This engagement requires a communication strategy starting with a CBA analysis to identify parties for the risk assessment, their manner of involvement, and the key messages to deliver to policy makers and different stakeholder interests. The communication strategy should include:

- An advocacy component that seeks, where needed, to remove the stigma of talking about corruption
- Awareness raising of corruption impacts, showing clearly the common interest in combating corruption (see table 1.7 in chapter 1).
- An effort to seek consensus on a framework of measures to prevent and detect corruption and to outline stakeholder roles for implementation.

What Makes a Communication Strategy Comprehensive?

Three steps in the process of formulating a communication strategy help task managers ensure that the strategy is sufficiently comprehensive and flexible:

1. Reflect in the strategy all dimensions of modern communication practice contextually: development communication, internal communication, corporate communication, and advocacy communication.

Twenty years ago, the World Bank did not talk about corruption. The staff called it the c-word, our shareholders and Board said it was too political, and we self-censored it out of our documents. Today fighting corruption is a key part of World Bank projects and programs. Our shareholders know corruption is a drag on economies, taxes the poor, and strangles opportunity.

An empowered public is the foundation for a stronger society, more effective government, and a more successful state.

—Robert B. Zoellick, President, The World Bank Group
address, the Peterson Institute for International Economics, April 6, 2011

2. Prepare the strategy openly and inclusively with ownership by stakeholders, because genuine involvement will ensure a sufficiently comprehensive, relevant, and value-added plan.

3. Prepare a new or updated project communication strategy to start each new phase of the infrastructure project cycle (i.e., while moving from planning and design to implementation and operation) to make improvements as the cycle progresses.

How Are These Steps Reflected in Practice?

In the Berg Water Project in South Africa, the Department of Water Affairs and Forestry (DWAF) was responsible for the communication strategy during the strategic planning, project identification, and approval stages of the BWP. During that period, the DWAF comprehensively expanded the communication strategy for the strategic demand-supply reconciliation studies and project environmental impact assessment (EIA) to reflect new water legislation and the "co-operative governance" model embodied in the constitution (1996).

Once construction was authorized, the TCTA assumed responsibility for developing the communication strategy for the implementation stage. It did this in cooperation with its major institutional partners, DWAF and the City of Cape Town, and residents of the Berg River basin. Box 2.6 describes the three interrelated streams of the implementation phase communication that the TCTA prepared in 2003.

The public participation process stream formed during the approval stage reflected statutory requirements to involve stakeholders in issues vital to maintaining public support. This involvement included government commitments to linking decisions on the physical design and operation of the project to the Berg River Catchment Management Strategy and Environmental Reserve Determination for the river system.

The PPP communication strategy was also benchmarked against the DWAF's Generic Public Participation Guidelines (containing 16 principles) and the recommendations of the World Commission on Dams on stakeholder involvement.[11] As custodian and facilitator of open dialogue processes, the PPP communication stream consolidated the role of the implementing agen-

Box 2.6. Communication Strategy: Three Streams for the BWP Implementation Phase

Public participation process (PPP). Facilitates communications integral to stakeholder engagement in ongoing decisions about design and implementation of the project, largely organized around the environmental management plan and social aspects.

Environmental monitoring committee (EMC). Facilitates functioning of the multistakeholder EMC as a cooperative governance mechanism and as a two-way dialogue between EMC members and their constituencies.

Project communication. Disseminates timely, accurate information on project components, status, and implementation issues to the public and stakeholder interests via newsletters, media, and various channels.

cies in managing relations with various public, civil society, and private sector interests.

The EMC communication stream was precedent-setting for South Africa. It enabled the multistakeholder EMC to have a legally recognized, collective voice and capacity for independent public and media communications.

With its substantive communication protocol, the EMC formulated its communication strategy around the environmental monitoring and management plan and social sustainability components of the project.

Consistent Messages as Part of a Comprehensive Strategy

Consistent information and messages from the major institutional partners are important. The BWP placed great importance on this goal because of the complex approval history and range of interests arrayed around implementation and operation decisions. The TCTA's role was accepted as the "first among equals" in coordinating government communication on the project but not to the point where the coordination biased outcomes.

The TCTA developed a protocol to which the other main institutional partners agreed. The terms of the agreement captured in the communication strategy included the following:

- Partners convey consistent core messages to the public and their respective stakeholders. The main message was that the Berg Water Project was an effective partnership between TCTA, DWAF, and CCT that delivered on its promises: a high-quality water supply project, associated social development, ongoing demand management, as well as integrated water resource management.
- Partners adapt core messages to various audiences, including information about delays, problems, and failures.
- Partners participate in public announcements through the media, advertisements of major events, sponsorships of surveys, participation in national campaigns, or any other public communication process deemed beneficial.

Table 2.4. Some Dos and Don'ts in Formulating a Comprehensive Communication Strategy

Don't do this	Do this
✄ Don't assume the communication specialist can produce the project communication strategy in isolation.	✓ Do base the communication strategy on a baseline assessment and a strong analysis to deliver project objectives and stakeholder value.
✄ Don't forget to benchmark the comprehensiveness of the communication strategy and use the results in developing further strategy.	✓ Do prepare a concise overall project communication strategy and ensure that each governance and sustainability aspect has a clear plan with the appropriate detail.
✄ Don't forget that ideally each major project partnership needs a communication strategy and a protocol for member communication.	✓ Do ensure that multistakeholder bodies can contribute to the communication strategy and take their comments into account.
✄ Don't assume that engagement of the media can be avoided and that a clear strategy and messages for the media are unnecessary.	✓ Do align the project communication strategy with the corporate communication strategies of the implementing agency and major institutional partners.
✄ Don't forget to include communication capacity building in the communication strategy.	✓ Do link communication objectives to the objectives of the project and broader sector policy on governance and sustainability improvement.
✄ Don't forget to include communication with the construction workforce in the strategy.	✓ Do prepare communication strategies and communication activity plans for all multistakeholder bodies on the project.
✄ Don't forget to formulate a new communication strategy for the start of each stage of the infrastructure project cycle.	✓ Do consider communication protocols for the major institutional partners so they disseminate clear and consistent information and messages to all.
✄ Don't forget to update the project Web site monthly at a minimum with postings such as the minutes of key partnership meetings.	✓ Do establish a project Web site for national and external stakeholders, and do put the communication strategy on the Web site.

- Although some divergence in conveying messages could (and did) occur, in the event of a crisis a crisis communication plan would be drafted and managed by the TCTA.

In addition, the communication strategy had to complement communication for other key sector programs. Important in this sense was (1) vertical integration with the sector communication strategy to implement policy and institutional reforms in water resource management and water service delivery and (2) horizontal integration with the communication strategies of other water programs and projects connected to BWP performance, such as the downstream water quality management programs.

Corporate Communication as Part of a Comprehensive Strategy

The corporate communication strategies of private companies and public enterprises are relevant to implementing dam projects, particularly in terms of feeding accurate and timely information to stakeholders and the media. In South Africa, relevance was illustrated by the TCTA linking the BWP project communication to its own corporate communication strategy through information and messages conveyed to shareholders via its Web site, media statements, and annual report.

The TCTA's corporate communication strategy promoted (1) its mandate within the countrywide reorganization of water service delivery to implement bulk water supply projects that conformed to international principles for sustainable infrastructure development and management, while recovering capital and operating costs from tariffs; (2) the alignment of the bulk water supply business with the relevant public policy, including cooperative governance, access to information, and black empowerment; and (3) its emphasis on corporate social responsibility, anticorruption, and the philosophy of implementing water projects in partnership arrangements.[12]

Multiple Objectives from Key Messages to Advocacy

To be fully comprehensive, communications strategies must support innovation and leadership to develop practices on a range of issues:

- Demystifying the media process so the media are an opportunity rather than a threat. Demystification means taking steps to overcome the wariness of many practitioners to engage the media and to build good relations with key audiences through the media.
- Mobilizing stakeholder support for advocacy positions on governance and sustainability improvements that carry weight and have political relevance.
- Supporting functional partnerships, ensuring adequate representation and a level playing field, appointing designated spokespersons, controlling rumors, and ensuring that communication vital to partners is timely, open, and accurate.

Table 2.4 is a list of dos and don'ts on formulating a comprehensive communication strategy for dam projects.

Improving Communication along the Project Cycle

This section explains what improving communication along the project cycle means on Bank-supported dam projects, and it offers practical "how-to" ideas for task managers.[13] Two project cycles are of concern to practitioners, the World Bank project cycle and the infrastructure project cycle.

World Bank Project Cycle

This cycle, illustrated in figure 2.3, is similar to the project cycle that other international public institutions use for investment and development policy lending, with the possible exception of the macro planning stages. This handbook simplifies the Bank project cycle into four stages:

1. Macro planning—that is, a country assistance strategy (CAS) and poverty reduction strategy paper (PRSP)—and project identification
2. Project preparation (combining preparation, appraisal, negotiation, and approval)

Figure 2.3. World Bank Project Cycle for Investment Lending

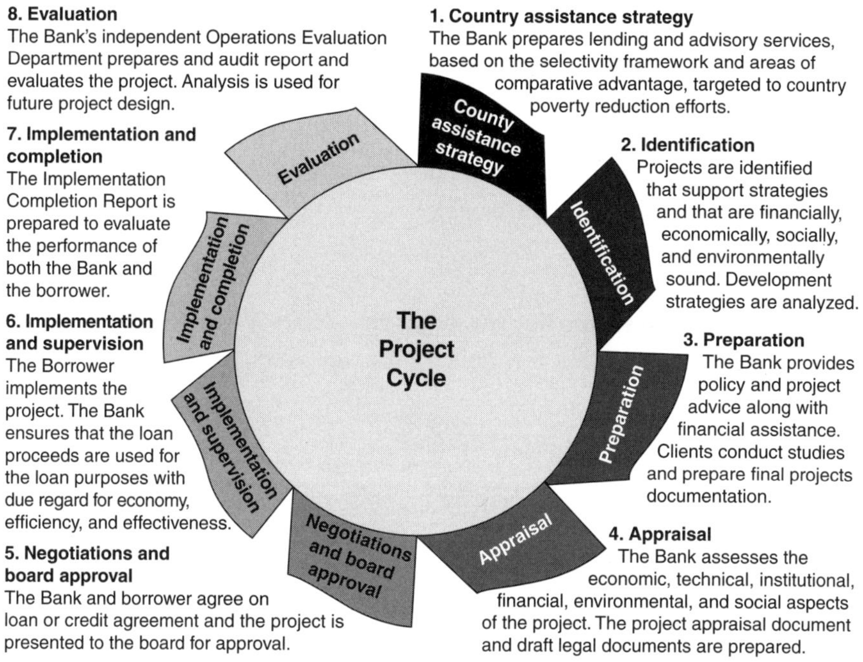

8. Evaluation
The Bank's independent Operations Evaluation Department prepares and audit report and evaluates the project. Analysis is used for future project design.

7. Implementation and completion
The Implementation Completion Report is prepared to evaluate the performance of both the Bank and the borrower.

6. Implementation and supervision
The Borrower implements the project. The Bank ensures that the loan proceeds are used for the loan purposes with due regard for economy, efficiency, and effectiveness.

5. Negotiations and board approval
The Bank and borrower agree on loan or credit agreement and the project is presented to the board for approval.

1. Country assistance strategy
The Bank prepares lending and advisory services, based on the selectivity framework and areas of comparative advantage, targeted to country poverty reduction efforts.

2. Identification
Projects are identified that support strategies and that are financially, economically, socially, and environmentally sound. Development strategies are analyzed.

3. Preparation
The Bank provides policy and project advice along with financial assistance. Clients conduct studies and prepare final projects documentation.

4. Appraisal
The Bank assesses the economic, technical, institutional, financial, environmental, and social aspects of the project. The project appraisal document and draft legal documents are prepared.

The Project Cycle

Source: World Bank.

3. Project implementation (implementation and supervision)
4. Project evaluation (combining implementation, completion, and evaluation).

The time pressures that task managers face in the project preparation and appraisal phases are a key consideration in "what is feasible" to improve practice relative to other stages of the project cycle. Part of the solution is creative consideration of how to deliver continual improvements in practices along the project cycle without simply loading more tasks into project preparation.

The Infrastructure Project Cycle

Non-Bank practitioners are more familiar with the infrastructure project cycle. The four main stages are illustrated in figure 2.4. Dam rehabilitation is highlighted as part of the operation stage because many dam projects in the World Bank portfolio involve rehabilitation, uprating, or modification of reservoir operations to conform to reforms for integrated water resource management (IWRM).

Consideration of both the Bank and infrastructure project cycles helps to define the best entry point to improve communication, governance, and sustainability practices contextually. In fact, there are multiple entry points, recognizing that Bank projects typically have three components: the first to support sector policy and regulatory improvements, the second for investment in hard infrastructure, and the third for capacity building.

Figure 2.4. Communication and the Infrastructure Project Cycle for Dam Projects

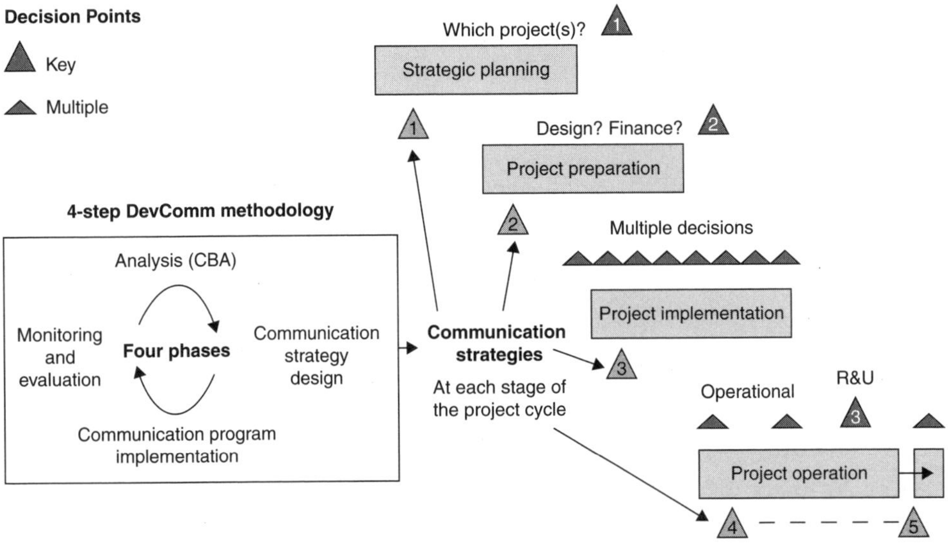

Source: Lawrence J. M. Haas.

The communication strategy must inform key decisions at each stage of the project cycle, providing the relevant communication analysis and implementation of the communication program to underpin public participation and stakeholder involvement in the most important decisions at each stage.[14]

Good practice today involves preparing a new communication strategy to move from one stage of the infrastructure project cycle to the next, using the four-phase development communication methodology depicted in figure 2.4. Why?

- *Actors change.* Institutional actors frequently change between stages. If all do not change, many institutional actors and stakeholders will perform different roles.
- *Risks and issues change.* Issues, risks, and stakeholder expectations change between stages, reflecting the nature of activities.
- *The policy environment evolves.* Time frames for dam projects are long: the implementation phase can span several years or more. Over this period, those in charge must take into account policy transformations, look afresh at opportunities for governance improvement, and rebalance expectations with social, environment, and economic performances.

Clearly, the communication strategy is built and adapted from one stage to the next. The case studies in appendixes A and B demonstrate this progression clearly.

The evolution of practice reinforces the need to build communication capacity within the implementation and operation agencies, the capacity for communication in multistakeholder partnerships that serve affected people in decision making, and especially the capacity and skills of the in-country communication specialists attached to the project who facilitate, assist, and train others in communication. On Bank-supported projects that have a wider development remit, the role and capacity of the local media to report on advances in governance and sustainability should be assessed.

The OECD's Development Assistance Committee (DAC) summed up the needs when in 2006 it called on communication and infrastructure practitioners to collaborate "to develop an effective communication strategy that covers all phases of the project" to advance "noteworthy achievements in quality, cost-effectiveness, and sustainability, as well as any incidents of alleged collusion, fraud, or corruption."

The World Bank Project Cycle for Investment Lending

This section looks at how interventions and products that reinforce communication, governance, and sustainability for dam projects map onto the Bank project cycle. It suggests the steps that task managers can consider in encouraging this three-way synergy. Chapter 3 provides more details on the specific tools.

Mapping Interventions and Products Along the Bank Project Cycle

Figure 2.5 illustrates how interventions and products in the four-stage development communication methodology map onto the Bank project cycle.[15]

What does such an approach mean for task managers of Bank dam projects? They recognize the following:

• More emphasis can and should be placed on macro- and sector-level communication analysis (a CBA) where policy formulation, strategic planning, and project identification take place.

Figure 2.5. Mapping of Development Communication Interventions and Products along the Bank Project Cycle

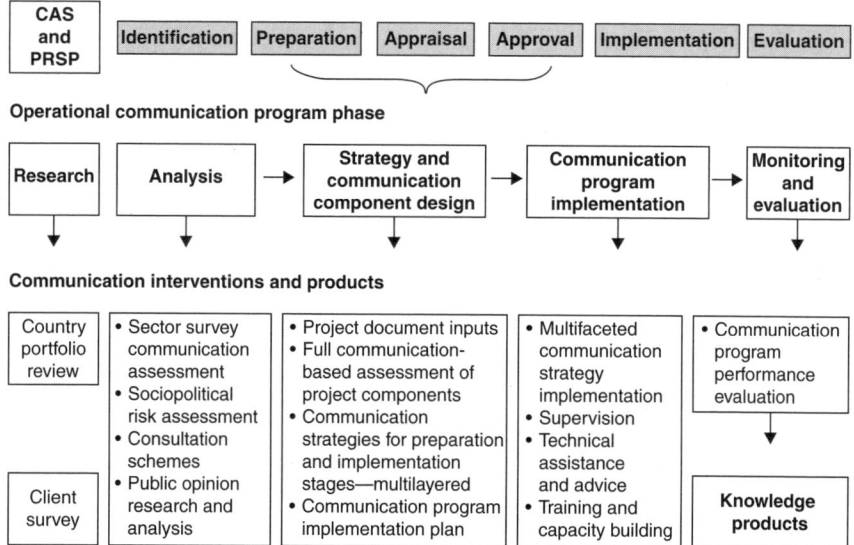

Note: CAS = country assistance strategy; PRSP = poverty reduction strategy paper.

- With project preparation and appraisal stages compressed in time, project communication activities must begin early to engage stakeholders effectively from the start. Otherwise, risks become stored-up problems.
- Because implementation of dam projects spans several years or more, communication capacity building should begin at the project preparation stage or early in the implementation stage. It is not capacity building after the work is done.
- The communication strategy for the implementation stage must be flexible and regularly revised. This strategy is more than just fine-tuning or reallocating within the communication resources.
- Project supervision budgets need a degree of flexibility to accommodate the introduction of good practice in implementation appropriately. Task managers need Bank-wide support to build contingencies into project supervision budgets that deliver continual improvement.

Using the same framework, figure 2.6 illustrates how the main governance and anticorruption (GAC) interventions and products relevant to dam projects map along the Bank project cycle.

Figure 2.6. Mapping of Governance and Anticorruption (GAC) Interventions and Products along the Bank Project Cycle

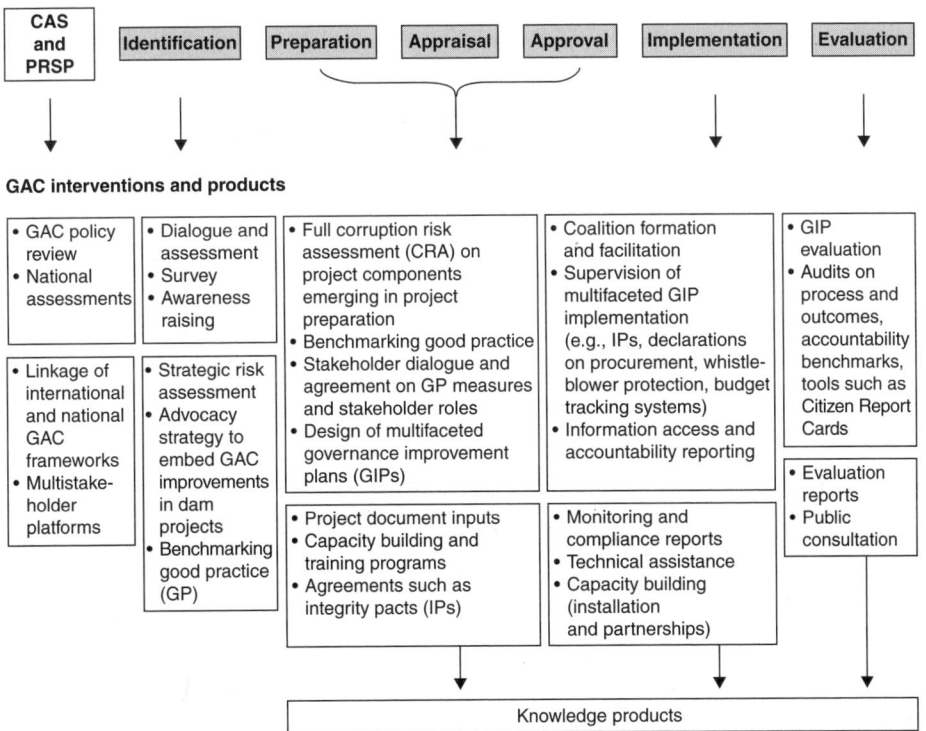

Note: CAS = country assistance strategy; PRSP = poverty reduction strategy paper.

What does such an approach mean for task managers of Bank dam projects? They recognize the following:

- It reinforces DAC's call for improvements in communication at all stages of the project cycle, recognizing that most if not all GAC tools have roles for stakeholders and measures to be taken at each phase.
- It reinforces the emphasis on governance-relevant analysis and assessment at the macro and sector levels by introducing GAC concepts into project identification documents and mobilizing the awareness and political will to incorporate GAC measures in the project design.
- It highlights that corruption risk assessments (CRAs) need to be part of an integrated approach to risk assessment and management, with potential efficiencies between sector-level and project-level CRAs. For example, a sector CRA would routinely and efficiently incorporate CRAs in all dam projects that move from planning to implementation stages.
- It shows that CRAs at the project preparation and appraisal stages contribute to the design of a governance improvement plan for the implementation stage, which encompasses a variety of tools for detecting and preventing corruption (identified in chapter 3).
- Finally, it shows the scope to incorporate stakeholder views of GAC in the project evaluation stage to capture perceptions of corruption to that point and perceptions of risk going forward, and to draw lessons for improving the policy framework and future project implementation.

Figure 2.7 illustrates how sustainability enhancement interventions and products map along the Bank project cycle. These actions improve the environmental, social, or economic performance of dam projects beyond the current environmental and social safeguards. This approach relates to the good practice the Bank promotes (such as an environmental flow assessment and benefit sharing) that are not safeguard requirements.

What does this approach mean for task managers of Bank dam projects? They recognize the following:

- The mapping reinforces that strategic (integrated) sustainability assessments undertaken during project identification and preparation stages are central to systematically identifying opportunities for introducing improvements in good practice sustainability.
- Existing strategic environmental assessment (SEA) tools can identify the relevant practices in collaborative processes. This objective and task, however, must be made explicit in the terms of reference for the SEA.[16] An integrated sustainability assessment in project preparation that looks at environmental, social, and economic performance holistically can reveal new opportunities and mobilize stakeholder awareness and support.
- It is important to identify the sustainability improvement measures that stakeholders prefer. Measures needing attention should be defined in detail and negotiated during project preparation. Those that can wait for further study should

Figure 2.7. Mapping of Sustainability Interventions and Products along the Bank Project Cycle

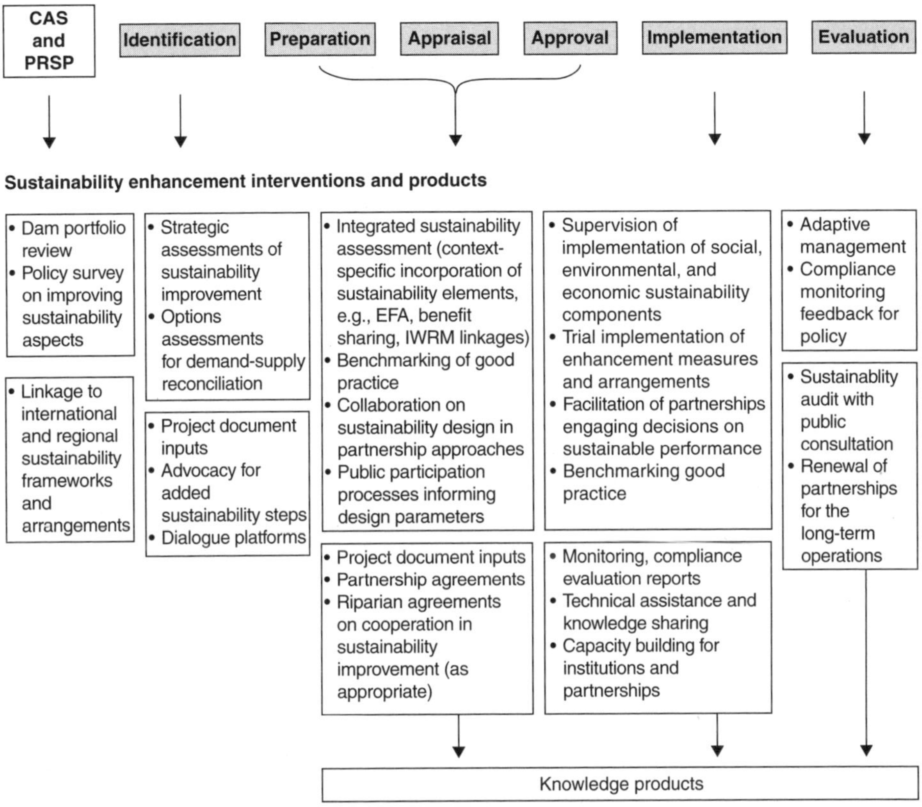

Note: CAS = country assistance strategy; PRSP = poverty reduction strategy paper.

be held for project implementation. Measures that would take effect after project commissioning, such as adaptive management of reservoirs, can wait.[17]

- The trade-off is that task managers must first identify a program for continual improvement in which measures could be introduced during project implementation. Second, and perhaps more important, at the preparation and appraisal stages task managers must remain vigilant to maintain the operational flexibility needed to accommodate the good practice in the implementation and operation stages that is provided for in the project design.

- Operational flexibility helps to rebalance the environmental, social, and economic performance aspects of the dam. This flexibility is found not only in the physical structures of the project (e.g., the size of outlets to accommodate larger downstream releases), but also in the various project agreements and legal documents. It is in the Bank lending instruments, as well as in the key agreements such as the power purchase agreements, bulk water supply agreements, and concession agreements. This multilayered flexibility provides scope for future operational changes and adaptive management.[18]

- This approach underlines the need to plan and deliver capacity building in the two- to five-year period after commissioning a dam project and to look for more creative ways to ensure adequate operation budgets to support stakeholders' roles in monitoring and adaptive management.[19]
- Evaluation-stage sustainability audits can contribute to discussions on introducing sustainability improvements for dams in the pipeline. This approach helps regulators to promote rather than constrain the adaptive management of dams.

Macro Planning and Project Identification

The CAS and PRSP are produced in collaborative processes with government and interested stakeholders. Initially, task managers become involved in project identification to generate the internal Project Concept Note (PCN). This process reflects macro planning priorities and sector analysis; it is where the need for a dam is identified. Figure 2.8 illustrates interventions and products that form the workspace to link the various activities that advance communication, governance, and sustainability at these stages.

To better capture three-way synergy, task managers may seek to ensure that the internal PCN and the Project Information Document (PID) released to the

Figure 2.8. Workspace in the Macro and Project Identification Stages

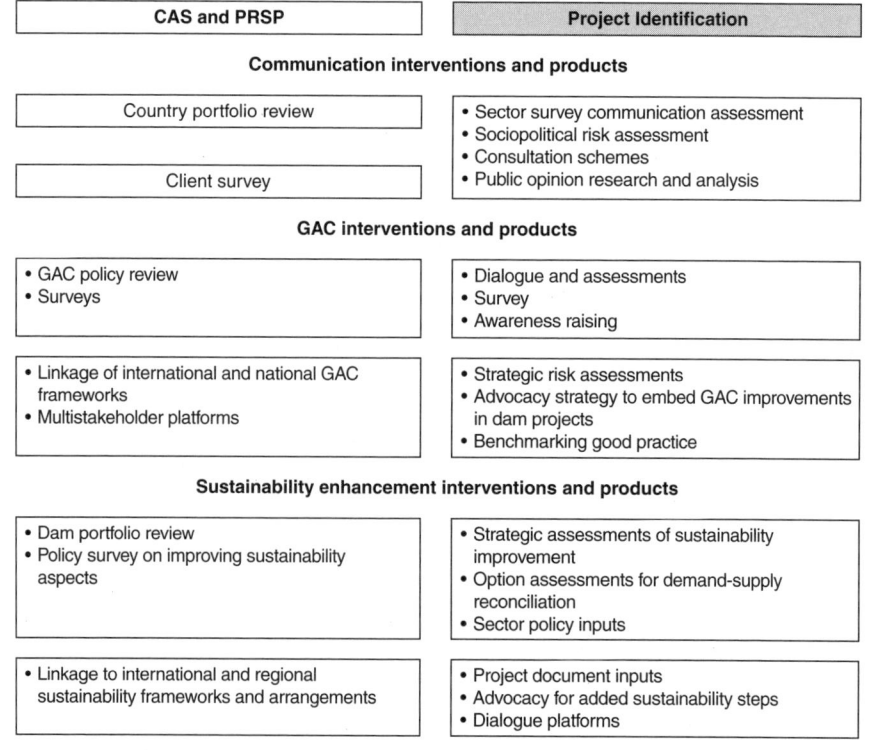

Note: CAS = country assistance strategy; PRSP = poverty reduction strategy paper; GAC = governance and anticorruption.

public reflect ongoing improvements in relevant practices and benchmarking, and that the Integrated Safeguards Data Sheet, a public document that sets out project preparation priorities, shows that governance and anticorruption are an integral part of project risk. Supporting analysis and technical advice can identify expectations and clarify stakeholder roles in connection with improvements in governance and sustainability on the proposed project, thereby ensuring that this philosophy moves forward during project preparation.[20]

More broadly, governments, CSOs, companies, and communities themselves have a role to play in advocating measures that meet their expectations so that projects can perform as development catalysts.

Making the business case for good practice is essential. Dam developers recognize it is increasingly difficult to do business without building good relations with communities and other stakeholders on issues of governance and sustainability.[21] Sector strategic assessments performed in open multistakeholder processes must enlist developers' support for good practice.

Project Preparation Stage   At the preparation stage, task managers play an advisory role and supervise the technical and financial support extended to the government and its implementing agencies. Although the Bank is wholly responsible for appraisal, it must provide stakeholders with an opportunity to review preparation outcomes and resolve outstanding questions on project design. The appraisal sets the tone for participation and partnership approaches going forward. Preparation can span several years. When the task team is under considerable time pressure, it is often compressed to two to three years (inclusive of project appraisal and approval).

Figure 2.9 depicts the interventions and products that form the workspace in which task managers have to advance activities for communication, governance, and sustainability. To better capture three-way synergy, task managers may seek to undertake the following:

- Ensure that government-led project preparation uses benchmarking and multistakeholder review to bring to light "good practice" for improvements in governance and sustainability.
- Advocate that a CRA is an integral part of the risk management framework and must be carried out in a collaborative way.
- Ensure that topic-specific assessments are not undertaken in isolation—for example, linking the CBA to the CRA and integrated sustainability assessments.
- Use stakeholder processes to inform decisions on which sustainability enhancement measures to detail in project preparation and which to detail closer to commissioning and operation.
- Review project design, agreements, and legal documents to ensure that they promote, not limit, the introduction of adaptive management.
- Involve community members on governing or multistakeholder advisory bodies as contextually appropriate.

Figure 2.9. Workspace in the Project Preparation Stage

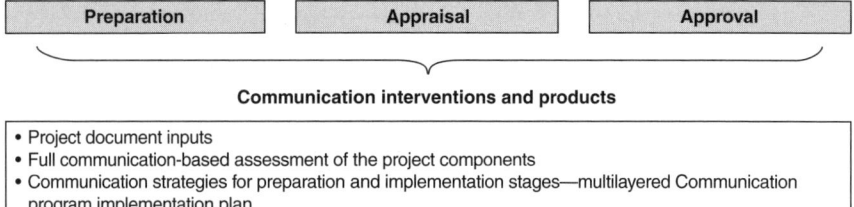

| Preparation | Appraisal | Approval |

Communication interventions and products

- Project document inputs
- Full communication-based assessment of the project components
- Communication strategies for preparation and implementation stages—multilayered Communication program implementation plan

Governance and anticorruption interventions and products

- Full corruption risk assessment (CRA) of project components emerging in project preparation
- Benchmarking good practice
- Stakeholder dialogue and agreement on good practice measures and stakeholder roles
- Design of multifaceted governance improvement plans

- Project document inputs
- Capacity building and training programs
- Agreements such as integrity pacts (IPs)

Sustainability enhancement interventions and products

- Integrated sustainability assessment:
  context-specific incorporation of sustainability improvements such as an environmental flow assessment (EFA), benefit sharing, integrated water resource management (IWRM) linkages
- Benchmarking of good practice
- Collaboration on sustainability design in partnership approaches
- Public participation processes informing design parameters

- Project document inputs
- Partnership agreements
- Riparian agreements on cooperation in sustainability improvement (as appropriate)

These steps are more likely to happen if the government and implementing agency take an active interest in the philosophy of continual improvement, as noted in the case study in appendix A.[22] Bank task managers can measure the commitment to continual improvement using the communication-based assessment.

Implementation and Supervision Stage   The borrower government implements dam projects over a period of years; the task manager is responsible for advice and Bank supervision. Implementation may continue for up to five years after the dam is commissioned—a period revealing sustainable performance. Although good practice improvements in sustainability can be introduced in the implementation and early operation stages (depending on their nature), improvements in governance to prevent and detect corruption in procurement are delivered during implementation.

Figure 2.10 illustrates interventions and products that form the workspace to advance aspects of communication, governance, and sustainability at the implementation stage.[23]

To better capture three-way synergy in supervision, it is helpful for task managers to do the following:

- Follow through with preset benchmarks for continual improvement in evaluating and introducing good practice.

Figure 2.10. Workspace in the Project Implementation Stage

Implementation

Communication interventions and products

- Implementation of multifaceted communication strategy
- Supervision
- Technical assistance and advice
- Training and capacity building

Governance and anticorruption interventions and products

- Monitoring and compliance reports
- Technical assistance
- Capacity building (institutions and partnerships)

- Coalition formation and facilitation
- Supervision of multifaceted GIP implementation (e.g., integrity pacts and declarations on procurement, whistle-blower protection, budget tracking systems)
- Information access and accountability reporting

Sustainability enhancement interventions and products

- Supervision of implementation of social, environmental, and economic sustainability components
- Trial implementation of enhancement measures and arrangements
- Facilitation of partnerships engaging decisions on sustainable performance

- Monitoring, compliance evaluation reports
- Technical assistance and knowledge sharing
- Capacity building (institutions and partnerships)

- Monitor changes in the policy environment that will unlock opportunities to introduce further good practice, using multistakeholder partnerships to review and advocate their inclusion.
- Closely monitor how well key institutional and multistakeholder partnerships function and provide appropriate communication support, including conflict resolution.
- Build capacity for stakeholder participation in the operation stage of dam projects that is vital to maintaining trust in governance measures and in improvements to the evolving sector and river basin management (IWRM) policy environment.
- Seek and support opportunities to benchmark corporate governance and the communication practices of implementing and operating agencies.
- Prior to project commissioning, prepare a new communication strategy for the operation stage, recognizing that risks and stakeholder roles will change.
- More broadly, task managers should seek ongoing improvements in communication practices, which require flexibility in supervision budgets.

Completion and Evaluation Stages   At the end of the disbursement period, task managers identify in an implementation completion report (ICR) the accomplishments, problems, and lessons learned. In the evaluation phase, the Bank's independent Operations Evaluation Department (OED) may conduct

Figure 2.11. Workspace in the Project Evaluation Stage

Evaluation

Communication interventions and products

• Communication program performance evaluation

Governance and anticorruption interventions and products

• Governance improvement plan (GIP) evaluation • Audits on process and outcomes (accountability, benchmarks, tools such as Citizen Report Cards)

• Evaluation reports • Public consultation

Sustainability enhancement interventions and products

• Adaptive management • Compliance monitoring feedback for policy

• Sustainability audit with public consultation • Renewal of partnerships for the long-term operations

a separate audit or prepare a project performance report (PPR) to measure project outcomes and their sustainability against the original objectives. The Bank's Independent Evaluation Group (IEG) reviews the ICRs using standardized criteria and may also perform more rigorous independent impact assessments of social, environmental, or other aspects.

Figure 2.11 illustrates the interventions and products at the evaluation stage. To better capture three-way synergy, it is helpful for the ICR and audit process to do the following:

- Highlight successes and failures in reflecting the philosophy of continual improvement from project preparation through implementation
- Emphasize the extent to which major-country policy and institutional reforms influence improvements in project governance and sustainability and the degree of integration in the project to understand whether opportunities were seized or missed
- Highlight accomplishments, problems, and lessons learned in adopting good practice communication, governance, and sustainability
- Indicate the attitudes, values, and perceptions of stakeholders revealed through participatory processes and partnerships that facilitated ongoing improvement.

OED and IEG evaluations should highlight good practice in governance and sustainability improvements in the World Bank's own portfolio of dam projects. The IEG criteria may look for the project's degree of benchmarking and use of the Bank's own good practice, seek reasons why these were (or were not) adopted, and propose remedies in general project cycle activities where appropriate.

Figure 2.12 unites these aspects to provide an overall picture of the relationships.

Figure 2.12. Governance, Sustainability, and Communication in Interventions at All Stages of the Bank Project Cycle

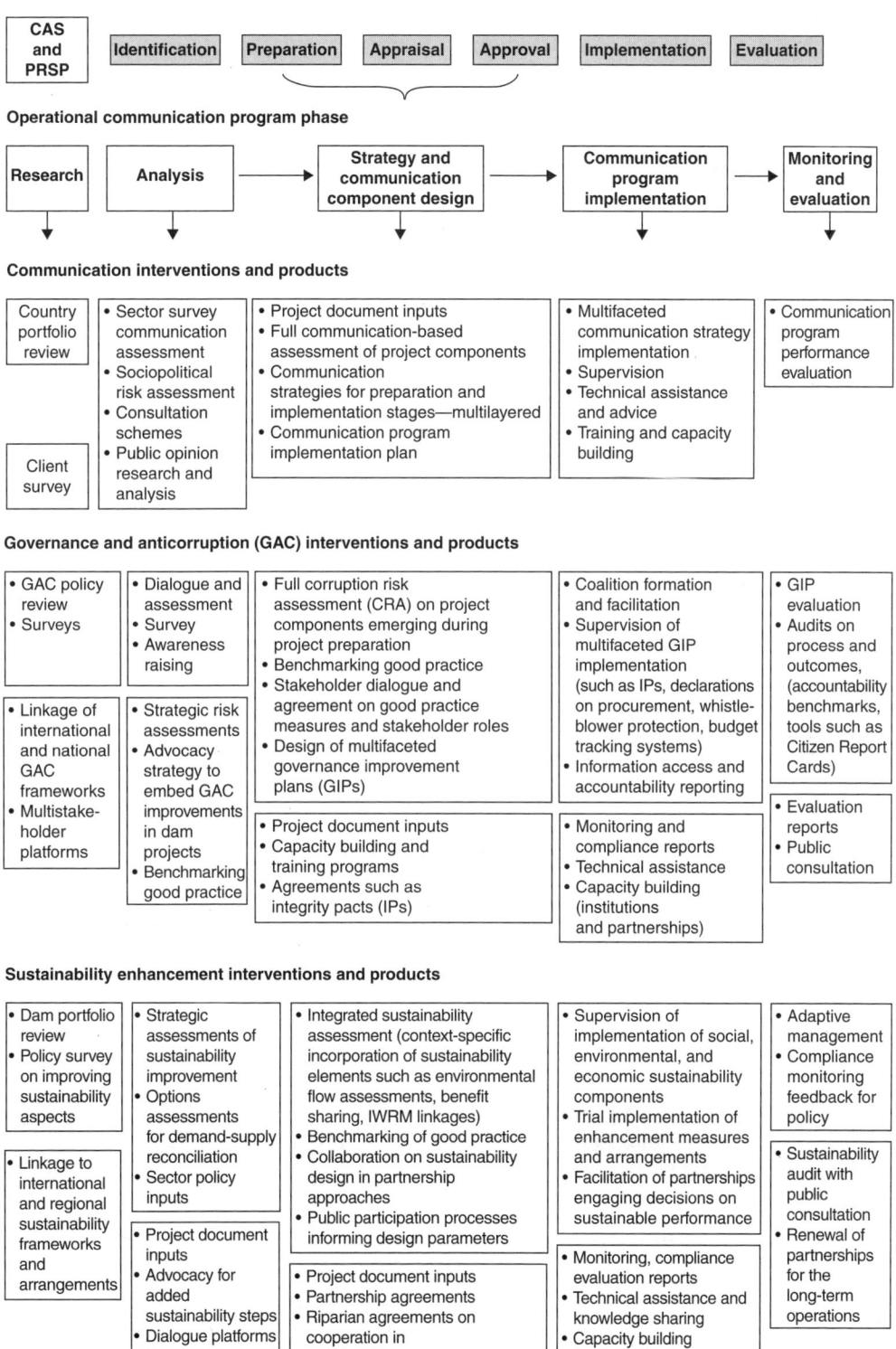

Table 2.5. Some Dos and Don'ts in Improving Communication along the Bank Project Cycle to Support Improving Governance and Sustainability in Dam Projects

Don't do this	Do this
✂ Don't forget that many dam developers now acknowledge their role to promote community development but have limited experience with community issues.	✓ Do introduce the philosophy of continual improvement at project conceptualization.
✂ Don't forget to monitor the functioning of key institutional and multistakeholder partnerships and their evolving communication support needs.	✓ Do share the Bank's good practice experience with government and project stakeholders from the start.
✂ In the identification stage, don't forget to embed the process of ongoing improvement.	✓ Do encourage working relationships and functional protocols for dam developers and local communities.
✂ In the preparation stage, don't forget to incorporate multistakeholder processes to benchmark against good practice.	✓ Do ensure there is adequate contingency and flexibility in communication budgets for project supervision.
✂ In the appraisal stage, don't forget to defer some aspects of good practice to implementation, but build intentions, flexibility, and triggers into project design.	✓ Do adopt good practices and supporting functional partnerships in implementation.
✂ In the implementation stage, don't forget to assess changes in the policy environment and to benchmark against good practice.	✓ Do look at the project agreements and legal documents at the appraisal stage to ensure they do not constrain recognized good practice and adaptive management, especially after project starts operation.

Table 2.5 is a summary of some dos and don'ts on improving practices and capturing three-way synergy during the conception and supervision of dams along the Bank's project cycle.

Investing in Communication for Governance and Sustainability

This section illustrates the budget implications of introducing good communication practices in support of improvements in governance and sustainability on dam projects. There is no simple answer to how much time and how many resources are involved. Two important questions, however, must be asked: Are project outcomes improved? How do good communication practices otherwise improve the capacity of practitioners to manage risks and stakeholder expectations?

Relevance for Communication Budgets

Table 2.6 depicts the relative impact of adopting good communication practices on communication expenditures at different stages of the dam project cycle.[24]

These improvements are to some extent reflected in today's trends in project activities. For example, figure 2.13 illustrates recent patterns in communication activities in Bank projects, though not exclusively dam projects. Although it looks at the basic question of communication from a different angle (communication activities that "seek inputs" and communication activities "that seek outputs"), it demonstrates how communication activities increasingly support stakeholder engagement processes.[25] However, the CBA remains underutilized.

Table 2.6. Relative Impact of Communication Improvements on Project Budgets

Stage of project cycle	Relative budget impact	Key improvements featured or delivered
Identification	None	Indicates a process of continual improvement over the project cycle
Preparation	Moderate to significant	Enhanced communication-based assessment (CBA)
		Enhanced communication support for stakeholder participation
		Enhanced governance and anticorruption (GAC) support for partnerships
		Introduction of good practice for improving sustainability
		Supervision/support for communication capacity development for operating entities
Appraisal	Moderate	Enhanced support for stakeholder participation in reviews
		Enhanced support for partnerships
Negotiation + APPROVAL	Modest	Agreement on a process of continual improvement over the project cycle: definition, milestones, triggers, etc.
Implementation	Significant Plus greater contingency and flexibility for ongoing improvement	Enhanced supervision/support for communication capacity of institutional and multistakeholder partnerships
		Supervision/support for four-stage preparation of operation stage communication strategy
		Supervision/support for communication capacity development for operating entities and stakeholder roles in adaptive management and operation stage governance improvement plan (GIP)
		Supervision/support for communication capacity development of media
Completion	Modest to moderate	Elicit stakeholder views and feedback in implementation completion report (ICR)
Evaluation	None to modest	Depends on Operations Evaluation Department (OED) and Independent Evaluation Group (IEG)

An example of this trend is the increase in spending on communication activities in the Lesotho Highlands Water Project from the Phase IA and IB project preparation stages to the current Phase II. Communication was a very small component of the LHWP total during the Phase IA and IB project preparation in the 1990s and early 2000s. By 2008, however, communication and public stakeholder consultation activities for Phase II feasibility studies accounted for more than 35 percent of the total budget.

Communications Budgets: Current Practice Baseline
Useful here is a look at the communication expenditures for two active Bank-supported dam projects: the 1,200-megawatt Nam Theun 2 project in Laos and the 50-megawatt Bumbuna hydropower project in postwar Sierra Leone.

The Nam Theun 2 project budgeted $250,000 to contract out the CBA activities and collaborative preparation of the communication strategy for the implementation phase.

Figure 2.13. Profile of Communication Activity on Bank-Supported Projects in 2007

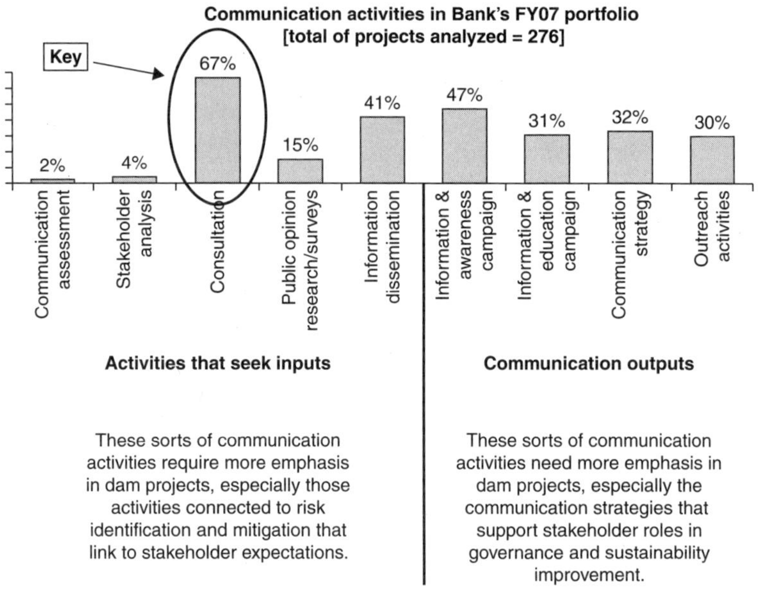

Communication activities in Bank's FY07 portfolio
[total of projects analyzed = 276]

Key →		Consultation 67%		Information dissemination 41%	Information & awareness campaign 47%			

Communication assessment 2%
Stakeholder analysis 4%
Public opinion research/surveys 15%
Information & education campaign 31%
Communication strategy 32%
Outreach activities 30%

Activities that seek inputs

These sorts of communication activities require more emphasis in dam projects, especially those activities connected to risk identification and mitigation that link to stakeholder expectations.

Communication outputs

These sorts of communication activities need more emphasis in dam projects, especially the communication strategies that support stakeholder roles in governance and sustainability improvement.

Source: Task Team Leaders Survey, Latin American and the Caribbean (LAC) Region, Development Communications Division, World Bank, 2004.

This figure, which excluded supervision costs, was just one item in the total project communication budget. The consultant hired provided assistance in conceptualizing, designing, and implementing a campaign, including a stakeholder mapping exercise, opinion research, and design and implementation of short- and medium-term communication strategies; holding training, workshops, and clinics; preparing effective public information materials; and developing and managing an interactive Web site.[26]

The Bumbuna project budgeted $200,000 for the CBA, development of a communication action plan during project preparation, and two years of implementation. This figure excluded supervision costs.

Incremental Costs of Improvements in Governance and Sustainability

In preparing this handbook, the authors estimated the incremental cost of developing a governance and anticorruption component of a water supply project. Table 2.7 shows the time and budget involved in project preparation.

The activities described in table 2. 8 revolve around a rigorous review of the corruption risks associated with the current project organizational structure and the development of a governance improvement plan to mitigate the main project corruption risks in implementation. It includes hiring a three-person team of experienced governance experts—international, regional, and local. The team would interview representatives of various project and stakeholder interests, including public sector, private sector, civil society, and funding organizations.

This example excludes the cost of a two-day workshop in which stakeholders would collectively review CRA results and discuss the main elements of the governance improvement plan (GIP) and their respective roles.

Table 2.9 outlines the content and cost of the workshop.

At the workshop, organizers would present the CRA and the outline of the GIP and the perspective papers by different stakeholder interests on the sought-after

Table 2.7. Communication Budget, Project Preparation, and Implementation Stages, Bumbuna Hydropower Project, Postwar Sierra Leone

A. Communication Action Plan	
– Salaries	$84,000
– Training	1,500
– Equipment	14,500
– Running costs	4,200
Communication with involved institutions	4,400
Communication with general public	47,778
Communication with local communities	38,500
Communication with international community (Web site)	7,840
Subtotal	**$202,718**

Table 2.8. Representative Cost of Corruption Risk Assessment (CRA) and Governance Improvement Plan (GIP) for a Water Supply Dam

Expected output	A plan setting out the sectoral/project corruption risk map, as well as a GIP
Orientation and linkage	Link to sector regulation and policy framework, looking at risk of corruption at all stages of the project cycle
Communication component	Parallel CBA analysis of attitudes toward corruption of key sector agencies and project stakeholder interests
	Preparation of awareness building and advocacy strategy
Estimated person-months	International, 2.0
	Regional, 1.0
	Local, 2.0
Estimated cost	$119,000

Table 2.9. Representative Cost of Multistakeholder Workshop for Three-Way Synergy from Governance, Sustainability, and Communication

Expected output	Recommendations by project participants of factors to emphasize in project preparation
	Stakeholder perspective position papers plus presentations
	Workshop dialogue and relationship building among stakeholders
Orientation and linkage	Participants would be major project stakeholders, including donors and representatives of local interests and ministries.
	Workshop could be organized as part of larger a stakeholder workshop using breakout sessions.
Communication component	Parallel CBA analysis of attitudes and expectations on corruption and the sustainability dimensions of the project
	Preparation of advocacy strategy and key messages for different stakeholders
Estimated person-months	International, 1.0
	Regional, 0.5
	Local, 1.0
Logistics	$10,000
Estimated cost	$70,000

improvements in sustainability and communication.[27] Outputs would include consensus on an action plan to use in the project preparation activities.

Project stakeholders (such as municipal representatives, affected community representatives, water user representatives, donors, concerned government agencies, procurement agencies, and civil society representatives) would be invited to participate in the workshop. Side events could include media and CSO capacity-building workshops. Breakout work group sessions would develop a priority list of steps that actors could take individually and collectively to build sustainability into the project, as well as the action and financing plans.

Notes

1. See Margaret Matters, *The Nuts and Bolts of Benchmarking* (North Carlton, Victoria, Australia: Alpha Publications Pty. Ltd., 1995).
2. Benchmarking is a broader management tool for improving the performance of organizations, regardless of the field or whether it is a public enterprise, private sector company, or civil society organization.
3. Transparency International provides specific suggestions on benchmarking anticorruption measures for Bank-supported infrastructure projects: http://www.transparency-usa.org.
4. Formulation of the strategy requires qualitative judgment, familiarity with good practice today, and access to the database on good practice that the Bank's Development Communication Division (DevComm) itself is continuously building though various activities, including its working paper series. Chapter 3 explains how non-Bank practitioners can access that series.
5. For example, the emphasis that should be placed in the benchmarking exercise on media relations and the media's roles, multistakeholder partnerships, crisis communication preparedness, the use of CBA tools such as opinion research, or issues dealing with the manner in which organizations are structured to respond to communication challenges and the work processes they employ in the management of the communication function.
6. As explained in the synopses in appendixes A and B and elaborated in the full case studies published by the World Bank, there were still challenges and failures. But most important, the mechanisms were in place. See Lawrence J. M. Haas, Leonardo Mazzei, Donal T. O'Leary, and Nigel Rossouw, *Berg Water Project: Communication Practices for Governance and Sustainability*, World Bank Working Paper No. 199 (Washington, DC: World Bank, 2010); and Lawrence J. M. Haas, Leonardo Mazzei, and Donal T. O'Leary, *Lesotho Highlands Water Project: Communication Practices for Governance and Sustainability*, World Bank Working Paper No. 200 (Washington, DC: World Bank, 2010).
7. As the Development Communication Division (DevComm) of the World Bank notes, neither general information nor targeted messages are absorbed, or shared, if stakeholders do not trust the government, project authorities, or others with whom they partner.
8. Combating corruption related to the project cost (which ultimately affects water and electricity tariffs) would, for example, add value for the low-income urban or periurban consumers most adversely affected by tariff increases. Similarly, the rural poor are often those most adversely affected by the failure to incorporate measures to promote environmental sustainability. Such a failure leads to the unnecessary loss of ecosystem services, which affects rural quality of life and livelihoods.
9. In an integrated risk management approach, "recognition of rights" and "assessment of risks" (particularly rights at risk) form the basis for more effective participation in handling risks in the planning and management of dams.

10. *King Report on Corporate Governance for South Africa.* Parktown, South Africa: Institute of Directors in Southern Africa, 2002.

11. As noted in the full BWP case study and in the interest of balance (in relation to pro-dam and anti-dam perspectives), some EMC members categorically rejected the PPP benchmarking, as well as several environmental NGOs and CSOs that had been affiliated with the Skuifraam Action Group (see appendix A).

12. The TCTA stated that its aim was to grow its reputation, not only to enhance collaboration with local partners on the BWP, but also to build public support for its involvement in future bulk water supply projects in the Western Cape and nationally. The TCTA's public messages explained that maintaining a good reputation helped it to maintain a good credit rating for projects. A good rating in turn helped reduce borrowing costs (thereby reducing water tariffs for consumers) and gave South Africa better access to financial markets for future bulk water supply investments.

13. As explained in chapter 1, the Word Bank's Development Committee called for infrastructure practitioners to improve communication along the project cycle to advance governance and anticorruption in Bank-supported infrastructure projects. This call is relevant to advancing sustainability in the provision of infrastructure.

14. The most important decision at the strategic planning stage is whether to advance and approve a dam project. Key decisions at the project preparation stage are related to the project design and financing arrangements. During project implementation, multiple decisions are made about the infrastructure, social, and environment management aspects. And during operation, key decisions are related to the adaptive management of economic, social, and environmental performance.

15. Adapted from an internal presentation, Development Communication Division, World Bank.

16. For example, to (1) incorporate a policy review, recognizing it is often possible to demonstrate that the legislation and existing policy framework calls for good practice sustainability improvement measures, but that the actual measures themselves have not yet been translated into dam projects; (2) benchmark against accepted good practice; and (3) provide a multistakeholder platform that enables discussion of priorities.

17. Task managers need not expend the time and resources to advocate the measures if the government does not see them as an immediate priority.

18. One example of this flexibility is how the bulk water supply agreement between the Department of Water Affairs and Forestry and the City of Cape Town incorporated the Berg Water Capital Charge that enabled the operator (TCTA) to make flexible changes in downstream releases from the dam to accommodate monitoring of water quality and ecological status without compromising its financial position and ability to service debt. This risk was borne by the consumers of water and energy service. It was wholly justified on the basis of internalizing in tariffs—not through a government subsidy—the cost of providing sustainable infrastructure.

19. Few Bank projects provide support for dam operations, but many aspects of the sustainable performance of dams hinge on adaptive management.

20. One example of strategic assessment in the governance area is Transparency International's Bribe Payers Index (BPI), which ranks the propensity of private enterprises in particular countries to pay bribes. TI's Corruption Perceptions Index (CPI) documents a country's reputation for honest practice. Countries with an adverse rating can be perceived as high on demand-side corruption.

21. Companies also recognize it is increasingly difficult to do business without building good relations with communities and other stakeholders. If companies understand that consultation or consent is part of the process of obtaining a legal license, gaining good community relations would be cheaper and easier. Companies are beginning to

acknowledge that they have a role to play in promoting community development, but many have limited experience with community issues.

22. Some developers will engage in efforts to improve governance and sustainability because of an expressed commitment to social responsibility and reputational interests. Others are required to do so by regulation and legislation.

23. Other practical aspects include ensuring that many of the task team members are not the same as those who participated in preparation of the previous project, and seeking a longer implementation phase, which will provide an opportunity to look at many new aspects of good practice. A key issue is to ensure flexibility to introduce relevant good practice over time.

24. Direct comparisons with past budget practices is difficult as it depends on how spending is now captured in project budgets—for example, whether growing expenditures for community consultation are captured in the resettlement or livelihood enhancement budget or the communication budget.

25. Activities that seek input are communication assessment, or stakeholder analysis, and public opinion research and consultations to inform decisions about and the design of project components. Activities that produce outputs are information dissemination, information and awareness activities, information and education campaigns, and outreach activities—all captured under communication strategies.

26. World Bank, *A Toolkit for Procurement of Communication Activities in World Bank-Financed Projects* (Washington, DC: World Bank, 2005), http://siteresources.world bank.org/EXTDEVCOMMENG/Resources/toolkitENfinal.pdf.

27. Whether this workshop is combined with review of benchmarking of the draft communication strategy would depend on factors such as the size of the project and complexity of those processes and whether a multistakeholder group had already taken ownership of the process to validate the project communication strategy.

Tools and Resources:
Some Thumbnail Sketches

Five Key Messages on Field Experience, Tools, and Resources

1. Many tools used for dam planning and management originate in the different areas of communication practice: development communication, corporate communication, internal communication, and advocacy communication. These tools are not used to their full potential in part because infrastructure practitioners are not aware of them.

2. The main tools used to embed governance and anticorruption measures in dam projects are (1) corruption risk assessments, prepared up-front in the project identification and preparation stages, and (2) governance improvement plans to implement recommended measures. Practitioners can choose from a wide variety of anticorruption tools. However, a consensus among stakeholders can help in choosing the most effective set for a particular situation.

3. Although the choice of tools depends on the context, an integrated sustainability assessment that considers all the dimensions of sustainability is a good starting point. This assessment, involving representative groups of stakeholders, ensures a common point of reference to determine how environmental, social, physical, financial, and institutional sustainability work together. For example, without financial and institutional sustainability any proposed measures for social and environmental sustainability become very difficult to implement.

4. Technical advice on how to apply these tools in specific contexts is available to World Bank task teams. The priority in applying the tools is to adopt collaborative approaches and to add value from stakeholder perspectives.

5. Anyone formulating training and capacity-building measures should look carefully at shared learning approaches for knowledge on mainstreaming tools. It is important for World Bank task teams to designate country counterparts for orientation training on these themes, including at a minimum one communication expert and one infrastructure practitioner.

Communication Tools

These thumbnail sketches of tools from the development communication field are written with infrastructure practitioners in mind, especially non-Bank practitioners who are less familiar with communication support from the Bank's Operational Communication Division.

This section begins with an inventory of communication tools, focusing on the four-phase development communication approach depicted here.

This inventory is followed by a section that describes at which stages of the World Bank and infrastructure project cycle the communication tools apply and add the most value.

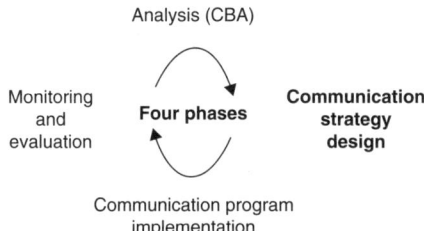

Development Communication Tools

Phase 1: Communication-Based Assessment (CBA)

- Communication needs assessment (CNA)
- Qualitative and quantitative communication research, including:
  - Polling
  - Perception studies
  - Baseline studies
  - Surveys
  - Interviews (client, project, sector, regional)
- Qualitative research (in-depth interviews and focus group discussions)
- Sociopolitical analysis
- Stakeholder analysis
- Participatory techniques
- Assessment of media and local communication capacity

The tools useful in phase 1 of the four-phase development communication approach are described in the sections that follow.

Communication needs assessment (CNA).   A CNA is a set of techniques used to assess local capacities, gain insight into sociopolitical concerns and roadblocks that affect the dam project, and determine the knowledge level and perceptions of the media about the project and related issues. It reveals what nongovernmental organizations (NGOs) are doing in the same area and verifies the availability and skills levels of research and communication agencies in the country. CNA studies rely on interviews and analysis of the available data to better understand the socioeconomic, cultural, and political contexts and to determine stakeholder perceptions, opinions, and beliefs.

Qualitative and quantitative communication research.   Based on qualitative and quantitative methods, this research delves into why people do what they do and think what they think. It investigates what behavioral or policy change is needed for the reform or development intervention to succeed. Ideally, this research links to the social assessments conducted in the framework of the project—that is, the project preparation studies. These research results help to decide on the directions for the communication policy, the improvements to be sought in operations, and the design of the communication strategy.

The specific activities performed include but are not limited to polling, perception studies, baseline studies, surveys, and qualitative research (in-depth interviews and focus group discussions).

Regional surveys inform the development of region-wide strategies and new products and services. These surveys can measure attitudes toward approaches among elites or the general population.

Sector surveys measure views on specific interests in a policy, project priority, or priority areas of development.

Project surveys inform the development of targeted communication strategies that resonate with critical audiences and the general population and provide teams with a more complete understanding of the project environment.

Client surveys are macro-level surveys undertaken in the development stages of the country assistance strategy (CAS) to engage with clients through a tracking survey sent to stakeholders in the country. This survey assesses the attitudes of stakeholders toward financing agencies and work in the country, as well as toward broader development issues.

Client satisfaction and value-added surveys are conducted for clients trying to measure their own clients' attitudes and sense of value added toward meetings, events, and so forth.

Sociopolitical analysis techniques.   Policies and development interventions cannot be effective without understanding the political environment not only for the initiatives but also for the communication functions that support them. This analysis provides details on the political risk environment for the reform policy or the project, the degree of transparency in similar development initiatives, the government's capacity to communicate with stakeholders, the institutional support for communications, the human resources needed to implement and evaluate communication strategies, and the legal framework for communication activities.

Various analytical tools such as public opinion polling are used to help decision makers in governments, development institutions, the private sector, and NGOs understand the stakeholder dynamics at the country, sector, and project levels. The analysis can also identify the pre-reform, political, or economic influences of change in society such as the media, NGOs, universities, think tanks, or ethnic and religious groups.

Stakeholder analysis techniques.   Stakeholder analysis is a tool for planning and managing relations on most development initiatives. It can assess not only stakeholders' degree of understanding and buy-in of the government's reform policy, program, or development initiative in any field, but also their perceptions and priorities. Identifying, disaggregating, and tracking stakeholders' perceptions and expectations lead to a better understanding of the opportunities and limitations of the project. With the support of the Bank's Operational Communication Division, the information can reveal segmentation of stakeholders and audiences, identify opinion leaders or allies, recognize the relevant social topics, and help task managers understand stakeholders' experiences and expectations.

Stakeholder analysis is commonly used in dam planning and management. However, the relative importance it is given by the task manager, and consequently the resources allocated to it, varies considerably. This variation affects its detail, scope, and usefulness to task and risk management.

Participatory techniques.   Today, most practitioners accept that the meaningful participation of stakeholders and partnerships from an early stage is central to informed decision making, leading to better-quality development initiatives and more sustainable results. Consultation implies the active solicitation of stakeholder perspectives to help shape projects and policy, instituting mechanisms for two-way information flow, building consensus, and providing conflict management when necessary. The CBA approach provides practitioners with the social mapping they need to advance constructive engagement with beneficiaries and interested parties who help to shape public attitudes from the beginning of a project.

Box 3.1 illustrates the practical questions addressed. These sample questions could relate to dam project management of resettlement and local development.

Techniques for assessing the media and local communication capacity. The Bank's Development Communication Division advocates two steps for determining the media tools appropriate for different target audiences: (1) identify the channels each segment of society uses to receive and disseminate information, and (2) evaluate the degree of trust existing in each channel.

Box 3.1. Social and Participatory Communication Analysis: Issues to Be Covered by the CBA

Social Forms of Information

- What are the spaces (such as schools and churches) of social interaction for the community?
- How are the beneficiaries represented? How are those leaders elected? Is the development initiative aware of these activities?
- How does the community make decisions (by means of meetings, committees about common problems)? How often? Have they discussed the topics related to the project in their meetings? What did they agree on?
- Have the beneficiaries produced some communication products in their communities? Can the project use these as communication mechanisms?
- Are there informal organizations in the community? What kinds of groups? What are their main activities?

Participation through a Two-way Communication Process

- Have the beneficiaries been consulted about the initiative? When? How were they informed? Which materials or messages did they receive? Which institution produced them? What was the final outcome of the consultation? What are the next steps for strengthening communication and interaction with them?
- How do the beneficiaries participate in the project? When did they start this process? What communication mechanisms do they use?
- Is the local political culture familiar with the concepts and outputs of the project?

Building Consensus

- Have the interests of the formal and informal community and local leaders been identified in project benefits and changes? What is the relationship among them? Have they worked together on past experiences? What were the results?
- Have there been previous conflicts among these people? Current conflicts? What is the magnitude? Could they affect the project?
- What common points have been identified for starting a consensus-building project? What is the degree of commitment for negotiating?
- Does the government support the community agreements? What is the community's opinion of governmental intervention? What about other stakeholders' interventions (such as the private sector or NGOs)?

Identifying Networks

- Who are the champions? How important are they for the beneficiaries and stakeholders?
- What other institutions are participating in community development? How do they coordinate to take actions? What is the level of community involvement? How do the other institutions measure this involvement?

The CBA framework media analysis provides a picture of different stakeholders' reasons for using specific communication channels, their level of trust in their sources, and the expectations generated by receiving information through mass media channels.

The tools useful in phase 2 of the four-phase development communication approach are described in the sections that follow.

Phase 2: Communication Strategy Design

- Definition of communication objectives
- Situation and risk analysis (ranging from political to impoverishment risk analysis)
- Detailed communication strategies and action plans

Definition of communication objectives.   This stage begins by finalizing and adjusting the communication objectives to the overall goals of the project or development initiative. This process is usually straightforward, based on the recommendations emerging from the CBA framework.

Table 3.1 outlines the sequential steps commonly taken in preparing the communication strategy, using the example of changing farmer behavior to control the incidence of fires in watersheds. This example is not specifically linked to dam planning and management or integrated water resources management (IWRM) in catchment management, but it does illustrate the technique.

Situation and risk analysis (ranging from political to impoverishment risk analysis).   This risk analysis is similar to the political risk analysis identified earlier in this chapter. This analysis, however, is able to focus more specifically on project and stakeholder risks, thereby complementing stakeholder analysis. Some tools in the social field, for example, cover risk analysis and so would work for project preparation studies and safeguard studies. From the communication perspective, this analysis identifies the most effective ways to deal with stakeholder concerns and perceptions about risk and to maximize the value that communication adds in risk mitigation for all stakeholders.

Communication strategies and action plans.   Community strategies and action plan documents guide communication implementation activities. A communication strategy is a well-planned series of actions aimed at achieving specific objectives through the use of communication methods, techniques, and approaches.

Table 3.1. Steps in the Design of a Communication Strategy

Basic steps	Main activities	Practical examples
1. Review the focal problem	Define the main problem and probe its causes	Fires started by farmers because of lack of knowledge of fire-control techniques
2. Define SMART objectives and identify key themes	Transform solutions into objectives, stated in a feasible and measurable way, identifying key messages	Reduce forest lost from uncontrolled fires by 80 percent within next two years
3. Define primary audiences/ stakeholders (1SHs)	Define and probe main group(s) of interest or define and probe audiences	Local farmers and their cultural and socioeconomic contexts
4. Identify secondary audiences/ stakeholders (2SHs)	Identify other groups of interest or audiences related to the overall objective	Farmers' families, other local actors, NGOs, government agencies
5. Define communication approaches and tactics	Assess objectives, audience context, and budget to select the most effective approach, linear or interactive mode	1SHs: capacity building, technical training 2SHs: awareness-raising media campaign
6. Select channels and media	Select most appropriate media for 1SHs and 2SHs	1SHs: use preferred sites and venues to provide information 2SHs: select appropriate media mix in that context
7. Design content topics and messages	Define key issues and most effective way to package them	1SHs: instructional design, key technical issues 2SHs: messages for raising awareness and knowledge
8. Assess the expected results of the strategic design for success of development initiative	1SHs: change in behaviors and practices to reduce forest destruction 2SHs: raising awareness of the importance of conservation	1SHs: farmers adopt more secure techniques to reduce incidence of uncontrolled fires 2SHs: become more aware of the importance and benefits of preserving forests

Source: Paolo Mefalopulos, Development Communication Division, World Bank (Sierra Leone), 2006.

Action plans set out the details (e.g., the design of communication materials and training of relevant staff) and instructions on producing and distributing media and information products. Figure 3.1 depicts the typical components of the communication action plan.[1] Typically, a dam project would include these components, as well as other components related to dealing with riparian countries.

An effective communication plan would ensure that all activities, costs, and responsibilities are defined, justified, and monitored.

The tools useful in phase 3 of the four-phase development communication approach are described in the sections that follow.

Figure 3.1. Components of a Communication Action Plan for Dam Projects

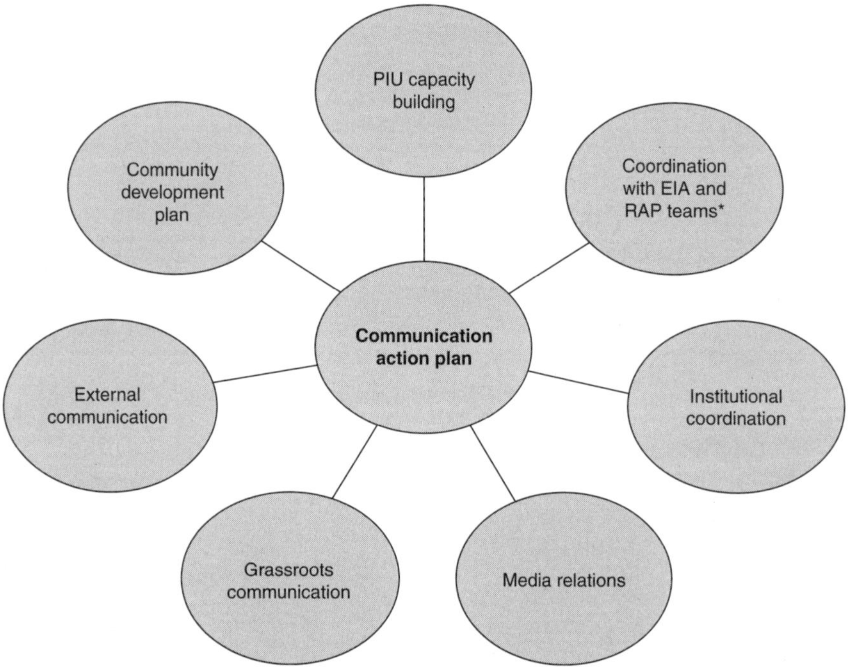

Source: Leonardo Mazzei.
Note: EIA = environmental impact assessment; RAP = resettlement action plan; PIU = project implementation unit.

Phase 3: Communication Implementation

Tools to implement activities specified in the communication strategy:
- Internal communication
- Corporate communication
- Advocacy communication
- Development communication

Decision template for five communication management objectives:
- Project-related information dissemination tools
- Conflict resolution tools
- Professional development and training
- Capacity-building tools

The communication action plan identifies the different communication tools and resources needed to implement the communication strategy.

Decision template for five communication management objectives. When communication management needs are relatively straightforward, the template for five communication management decisions can be used as a basis for monitoring the communication strategy for the project or program. The template essentially addresses five questions:

1. Which audiences need to be reached by the communication component?
2. What are the required behavior changes?
3. What messages are the most appropriate?
4. Which channels of communication will be most effective?
5. How will the communication process be monitored and evaluated?

In more complex projects, the template may be appropriate for aspects of the project that are relatively straightforward. For example, in environmental flow policy the template could be used in developing and monitoring communication support to contribute to reservoir operation parameters. Other aspects of the project, such as the public-private, may require more complex treatment.

Project-related information dissemination tools.   The following list of tools provided by the Bank's Operational Communication Division is by no means comprehensive.

- *Information dissemination/campaigns* are the targeted dissemination of information to fill specific knowledge gaps. This approach relies heavily on diffusion models through media campaigns, which apply in a number of circumstances, either for broader national audiences or for populations in specific areas. In the past, campaigns tended to rely heavily on a single, specific medium. Today, they take advantage of a mix of different media.
- *Information, education, and communication (IEC)* refers to a broader set of tactical approaches to disseminate information and to educate large audiences. It is based on the linear transmission model in which information is disseminated through a number of media.
- The *education/training approach* applies to programs requiring instructional design, often based on an interactive modality. Educational approaches aim to increase knowledge and comprehension, whereas training approaches focus on improving professional skills.
- *Institutional strengthening* seeks to strengthen the internal capacities of an institution (e.g., through training) or to position and improve its image with external audiences.
- The *community mobilization approach* is a systematic effort to involve the community in resolving specific issues related to its well-being. The approach may require the formation of groups designated to participate in the decision-making process and to follow up on specific issues (e.g., monitoring the activities in a project work plan).

The communication strategy will take into account the many channels that can be used to reach different audiences. These channels range from the electronic media, such as the Internet, print media and broadcast media, to the traditional channels, such as popular theater, drums, and storytelling.

Conflict resolution tools.   The CBA can advise on the different conflict resolution tools that may be needed to deal with issues arising among project stakeholders. These tools are well documented in the literature. What is important here is that the communication strategy identify the types of tools that can be deployed to support communications needs.

Professional development and training.   The existing formal, informal, and on-the-job training courses and programs are either available or can be accessed regionally. During implementation, the key stakeholders involved in the project would be given access to training based on needs identified in the CBA.

Capacity-building tools.   These tools fall into two categories: (1) tools to identify communication possibilities as part of a project or program, and (2) tools to build capacity for communications to support the project or the successful and sustained application of reform.

The tools useful in phase 4 of the four-stage development communication approach are described in the sections that follow.

Phase 4: Communication Evaluation

- Traditional results and ex post evaluations
- Participatory monitoring and evaluation
- Public opinion tracking studies
- Randomized design, baselines, and surveys

According to the Bank's Operational Communication Division, evaluation is a crucial aspect of development communication interventions. Measuring and evaluating the impact of social interventions are never simple, and in development communication that task becomes more difficult and complex, mostly because of its broad role. The following sections describe some of the tools used in communication-based monitoring and evaluation.

Traditional results and ex post evaluations (based on indicators of the ex ante evaluation).   Ex post evaluation tools and indicators can be developed for all communication interventions. Periodic feedback from the main stakeholders is of primary importance, however. This feedback may help redefine the implementation of both the development initiative and the communication program to better accomplish the development goals.

Participatory monitoring and evaluation.   This step is an extension of participation in the project design with the aim of informing decision making on all aspects of a project. A variety of tools and techniques can be used to involve stakeholders in determining the value and support offered by communication activities. For example, the implementation of community development initiatives with revenue-sharing funds and the implementation of environment and social mitigation management measures through local action can be highly communication-intensive activities (including, for example, local community radio and newsletters). Thus evaluation techniques are essential for ongoing improvements and for adapting management of the activities.

Public opinion tracking studies.   Tracking public views on issues such as public-private financing arrangements for hydropower and tariff reforms that ensure sustainable financing of environmental and social components is essential for the sustainability of the whole initiative.

Randomized design, baselines, and surveys.   A variety of survey techniques can calibrate and assess stakeholder attitudes on the high-value communication aspects of the project. Properly designed and implemented, these surveys can provide watershed managers and operators of dams with valuable information, as well as feedback for sector-based strategies to manage and improve the development performance of other dams in the country.

Operational support in the areas just described are available from the World Bank. The following are just some of the operational support activities offered by the Operational Communication Division, but they do indicate the type of support that the task managers and development practitioners responsible for hydropower projects and dam planning and management may require:

- Communication-Based Assessment
- Developing Communication Strategies
- Communication for Behavior Change
- Fostering the Participatory Process

Communication Tools along the Project Cycle

The World Bank's Development Committee (DC) recommended that task managers, to improve governance and counter corruption in infrastructure, improve communication strategies and tools at each stage of the project cycle. According to the DC, it is important to develop an effective communication strategy that covers all phases of the project. The communication plan must provide for consistent messages to be conveyed to all of the relevant stakeholders: government officials in the implementing agency; contractors, suppliers, and consultants who may be involved in bidding on the project; members of civil society affected by the project; and (as appropriate) the local press. The role of the media (and civil society) may be especially important if the plan includes the use of publicity—both positive and negative stories—as a tool for reducing the level of fraud and corruption in Bank projects. The objective would be to highlight both noteworthy achievements in quality, cost-effectiveness, and sustainability, as well as any incidents of alleged collusion, fraud, or corruption. Table 3.2 shows how to apply the communication tools

Table 3.2. Mapping of Development Communication Tools along the Project Cycle

	World Bank project cycle, stages					Infrastructure project cycle, stages				
	Macro planning (CAS)	Project identification	Project preparation, appraisal, approval	Project implementation	Project evaluation	Strategic planning	Project preparation and design	Construction/ implementation	Operation	Rehabilitation/uprating/ reoperation
Phase 1: Communication-based assessment										
Communication needs assessment	✓		✓				✓			
Qualitative and quantitative research: • Polling, perception studies, baseline studies, surveys • Interviews: client, project, sector, regional • Qualitative research: in-depth interviews and focus groups	✓	✓	✓	✓	✓	✓	✓	✓	✓	✓
Sociopolitical analysis	✓					✓	✓			
Stakeholder analysis			✓			✓	✓	✓	✓	✓
Participatory techniques	✓	✓	✓	✓	✓	✓	✓	✓	✓	✓
Media and local communication capacity assessment			✓				✓			
Phase 2: Communication strategy design										
Definition of communication objectives			✓							
Situation and risk analysis (ranging from political to impoverishment risk analysis)			✓				✓		✓	
Detailed communication strategies and action plans			✓				✓		✓	
Phase 3: Communication implementation										
Tools to implement activities specified in the communication strategy: • Internal communication • Corporate communication • Advocacy communication • Public communication		✓	✓	✓		✓	✓		✓	✓
Five communication management decisions template										
Project-related information dissemination tools			✓	✓			✓	✓	✓	

Table 3.2. *Continued*

	World Bank project cycle, stages					Infrastructure project cycle, stages				
	Macro planning (CAS)	Project identification	Project preparation, appraisal, approval	Project implementation	Project evaluation	Strategic planning	Project preparation and design	Construction/ implementation	Operation	Rehabilitation/uprating/ reoperation
Conflict resolution tools		✓	✓			✓	✓	✓		
Capacity-building tools		✓	✓			✓	✓	✓		
Professional development and training: formal, informal, on-job		✓	✓			✓	✓	✓		
Phase 4: Communication evaluation										
Traditional results and ex post evaluations (based on indicators of the ex ante evaluation)				✓						
Participatory monitoring and evaluation				✓						
Public opinion tracking studies				✓						
Randomized design, baselines, and surveys				✓						
Qualitative program and product evaluations				✓						

Note: A communication strategy would be prepared for each stage. This table indicates the main use of the tools.
CAS = country assistance strategy.

just described along both the Bank project cycle and the infrastructure project cycle, as discussed in chapter 2 of this handbook.

Each new stage requires a new communication strategy, repeating the four-phase process, and each successive strategy builds on the previous stage of the project. As issues change along the project cycle, the stakeholders' views of risk may change and the stakeholders themselves may change or operate in different capacities.

Anticorruption Tools

This section is a primer on the anticorruption tools and techniques relevant to dam planning and management. This description of these tools is adapted from a detailed discussion of the tools by Transparency International.[2]

This section begins with the tools needed to embed anticorruption measures in government reforms and development projects. It identifies the range of anticorruption tools, starting with the primary ones for dam projects, corruption risk assessments (CRAs) and governance improvement plans (GIPs). Both the CRA and GIP require a clearly articulated communication component that reveals the engagement of all stakeholders to ascertain their perceptions of corruption—its risk, type, impact, detection, and prevention.

Thumbnail sketches of the CRA and the GIP are followed by those of other anticorruption tools organized by use at the national, sector, and project levels. Some of these tools would be incorporated in a GIP for dam projects.

An illustrative mapping of these tools along the project cycle follows.

Types of Anticorruption Tools

Corruption Risk Assessment

A corruption risk assessment is an explicit analysis of the functional areas in a development project or program to uncover vulnerabilities to corruption. With stakeholder participation, the assessment identifies major corruption risks, assesses their potential impact, and determines the adequacy of the control environment, using analytical communication techniques for risk assessment.

Good practice suggests a structured process with full engagement of the range of stakeholders in the assessment of risks and public consultation on the results. Valuable outcomes include recommendations for additions to or modifications of an organization's operational practices, procedures, systems, or controls, so that the risk of corruption is reduced.

Governance Improvement Plans

A governance improvement plan is a specific plan of action to (1) assess corruption risks in a specific program or project; (2) identify anticorruption measures involving all stakeholders in a collaborative process; and (3) outline a strategy for systematic implementation and monitoring as an integral part of the project.

The GIP embodies a range of measures identified in the CRA process and multistakeholder review. It includes a communication strategy that incorporates two components: an advocacy component directed at policy makers and their role in creating policy and regulation and a public participation component to engage the public and to seek support of the necessary partnerships to implement specific anticorruption measures.

Infrastructure is classified as a sector at high risk for corruption. Although presently there are few examples of GIPs on hydropower projects, some elements of GIPs are at work. The road sector is perhaps the most advanced in providing examples of anticorruption and governance improvement plans, and they could be readily adapted to hydropower and dam planning and management.

Other Anticorruption Tools

To construct a more comprehensive approach, the fight against corruption in dams can be waged at the national, sector, or project level.

Anticorruption Tools Available at the National Level

- National integrity system (NIS)
- National integrity pact (NIP)
- Access-to-information laws
- Anticorruption laws
- Lobbyist registration
- Complaints and ombudsman office
- Independent anticorruption agencies
- Office for access to public information

The sections that follow describe examples of the currently available tools applicable at the national level.

National integrity system (NIS).   This system approach takes the "Institutional Pillars of Integrity" as a starting point and recognizes that societies become resistant to corruption when a whole range of institutions are present and function well. Organizations work together to develop good governance practices to fight corruption and take mutually reinforcing steps in their own spheres of influence or "shops."

The NIS identifies an explicit set of reinforcing measures that encourage public officials and companies to refrain from practices such as bribery. And it aims to give companies greater assurances that their competitors will also cooperate, so that those who adopt good governance practices are not penalized. Governments and government officials also have the assurance of a clear framework that protects them from dubious offers: they know clearly what form of behavior is not acceptable. The NIS introduces a monitoring system that provides for independent oversight and accountability, typically involving civil society.

In 2006 an NIS was prepared for Japan.[3] It looks at "the enormous public-works burden on Japanese local governments that renders these governments highly vulnerable to pressure for corruption in building dams." As a tool, such country studies provide a starting point for identifying areas that require priority action and a basis for stakeholders' assessments of existing anticorruption initiatives. In general, the NIS country studies help explain which pillars of the NIS have been more successful, why, whether they are mutually supportive, and what factors support or inhibit their effectiveness.

National integrity pact (NIP).   An NIP is a specific agreement reached between a government or a government department (at the federal, national, or local level) and bidders for a public contract. The agreement sets out the rights and obligations for all sides: (1) no payment, offer, demand, or acceptance of bribes; (2) no collusion with competitors to obtain the contract; and (3) no engagement in such abuses while bidding for, approving, negotiating, or carrying out the contract.

Access-to-information laws.   Most countries have legislation providing access to information. The extent to which these laws are applied in practice, however, can vary considerably.

Anticorruption laws.   Most countries have some form of criminal legislation related to corruption and bribery. The concern is to ensure specificity in the laws and the appropriate implementation mechanisms. Advocacy may center on introducing tighter and more specific anticorruption laws, enforcing the compliance of existing laws, or both.

Lobbyist registration.   Registering lobbyists is an effective tool for increasing the transparency around lobbying for specific projects and policies or lobbying to prevent new legislation that promotes transparency and accountability.

Complaints and ombudsman office.   This office enables the diverse voices of society to raise concerns about corrupt practices and to call for time-based actions to investigate and recommend measures to stamp out corruption in different sectors.

Independent anticorruption agencies.   These agencies provide independent oversight and capacity (with legal power) to investigate reported instances of corrupt behavior and practices at all levels, including political levels.

Office for access to public information.   This office establishes standards for public information and seeks to overcome problems with weak institutions and lack of resources. It makes critical documents on strategically important projects available to any interested party; members of the public, civil society, and the media can view official information at these offices. Corruption thus becomes more difficult to pursue and to conceal. A central office does not replace the need for project-specific information offices, particularly those in the project locations.

The sections that follow describe the currently available tools applicable at the sector level.

Business principles for countering bribery.   The principles are a practical tool for companies seeking a model or template to adopt a comprehensive

Anticorruption Tools Available at the Sector Level

- Business principles for countering bribery
- Minimum standards
- Codes of ethics
- Public service disclosure of income and assets
- Oversight committees
- Independent audit function
- Accountability benchmarks
- Surveying and monitoring perceptions of corruption

antibribery program and to find a starting point for developing their own antibribery programs or providing a benchmark for existing ones. From the industry or private sector perspective, it is important that infrastructure actors work together and with other stakeholders groups. Transparency International has witnessed many situations in which industry members have worked together.

Minimum standards.   Minimum standards can be developed and applied to needs assessment, design, preparation, and budgeting activities prior to the contracting process; to the contracting process itself; and to contract implementation. Standards can extend to all types of government contracts (including procurement of goods and services); supply, construction, and service contracts (including engineering, financial, economic, legal, and other consultancies); privatizations, concessions, and licensing; subcontracting processes; and involvement of agents and joint-venture partners.

Codes of ethics.   Codes of ethics or codes of conduct are important guides to making decisions on complicated ethical issues. They provide the basis for an environment in which citizens are aware of the basic standards of behavior expected from public sector employees. Codifying standards of behavior is not by itself, however, sufficient to ensure ethical conduct by public officials.

Public service disclosure of income and assets.   Many countries have rules requiring public officials to declare their assets and wealth because disclosure reduces the chance of corruption. The purpose of public officials' declarations is to identify what wealth is not fairly attributable to income, gift, or loan. The compilation demonstrates the wide range of approaches to restraining officials' undesirable conduct.[4]

Oversight committees.   Sector-level committees can provide oversight on issues as diverse as awarding licenses to private power developers to implementing codes of ethics within the agencies that implement public procurement.

Independent audit function.   Formal audits of public and private project agencies conducted by independent entities help to prevent and detect corruption. Public dissemination of audit information helps to build trust in the project. Prompt publication of audit reports ensures that their impact is undiminished.

Accountability benchmarks.   These benchmarks identify the main areas needing better accountability and formulate the questions at the core of the problem, allowing comparison of answers from public servants (and political levels) and monitoring of subsequent performance.

Surveying and monitoring perceptions of corruption.   Surveys at the national or sector level help to increase public awareness of the level of

Anticorruption Tools Available at the Project Level

- Governance improvement plans embedded in projects
- Corruption risk assessments
- Integrity pacts for public procurement
- Project anticorruption system (PACS)
- Minimum standards
- Codes of ethics
- Project compliance plans

corruption in various sectors, such as hydropower development in the water sector, and to understand views on how to tackle corruption.

The sections that follow describe the currently available tools applicable at the project level.

Governance improvement plans embedded in projects.   As detailed earlier, this specific plan of action (1) assesses corruption risks in a specific program or project, (2) identifies anticorruption measures involving all the major stakeholders in a collaborative process, and (3) sets out a governance improvement plan that will be systematically implemented and monitored as an integral part of the project.

Corruption risk assessments.   As detailed earlier, this is an explicit assessment of the risk for corruption in a development project or program, using a range of communication techniques. Good practice includes an independent assessment of risks, public consultation, and a structured process to fully involve the main project stakeholders.

Integrity pacts for public procurement.   An integrity pact revolves around an agreement between a government or a government department (at the federal, national, or local level) and all bidders for a public contract.[5] The agreement sets out rights and obligations for all sides: (1) no payment, offer, demand, or acceptance of bribes; (2) no collusion with competitors to obtain the contract; and (3) no engagement in such abuses while bidding for, approving, negotiating, or carrying out the contract. The integrity pact also introduces a monitoring system that provides for independent oversight and accountability.

Project anticorruption system (PACS).   PACS is an integrated, comprehensive system to help prevent corruption on construction projects. The system comprises PACS Standards, which are anticorruption measures for construction projects, and PACS Templates, which are the tools used to implement PACS Standards.

Although PACS is designed as a project system, some modules (such as disclosure, training, and rules for individuals) may be used by companies. PACS measures have impacts on all project phases, on all major participants, and at various contractual levels. PACS may be a prerequisite for project approval, or it may be required by financiers as part of the funding package or by public or private sector project owners as a condition of participation in a project.

Minimum standards.   Like minimum standards at the sectoral level, minimum standards at the project level could be a component of the governance improvement plan.

Codes of ethics.   As at the sector level, codes of ethics could apply not only to procurement activities during project construction but also to the operation stage, during monitoring and evaluation and during the provision of services.

Project compliance plans.   Plans for dam projects are recommended by the World Commission on Dams (WCD), identifying all the major commitments needed for the social and environmental components of the project and for monitoring. Compliance on all aspects is reviewed with the key stakeholders and in public consultations as appropriate.

The sections that follow describe examples of other anticorruption tools.

> **Selection of Other Anticorruption Tools**
> - Participatory budgeting and budget tracking systems
> - Citizen Report Cards as an aid to public accountability in service provision and access
> - Whistleblower protection
> - Public participation and education tools
> - Ethics training
> - Anticorruption radio spots and media campaigns

Participatory budgeting and budget tracking systems.   A budget or expenditure tracking system can be used to report transparently on expenditures for programs and projects. Whether simple or more complex, these systems must make the information accessible and open for scrutiny. In one of the many practical illustrations of how these systems help to detect and prevent corruption, in Sierra Leone a public expenditure tracking survey exposed overall discrepancies between money allocated to line ministries and money received at the grassroots level.[6] The different components of dam projects offer considerable scope for budget tracking, including the expenditure on resettlement and compensation.

Citizen Report Cards.   This tool enables the beneficiaries of programs or projects to report their exposure to corruption as an aid to public accountability in service provision and access. Indeed, Citizen Report Cards help to open channels of communication between governments and their citizens. In Mongolia, for example, hotlines have been established to provide citizens with a means of reporting administrative-level corruption that affects them directly. Citizens are also now contributing to the corruption report cards being generated from the calls into the hotlines. The initial results of this pilot effort were publicized at the Press Institute, and they contributed to the Asia Foundation's goal of advancing pride and integrity in the public and private sectors.

Whistleblower protection.   This protection is extended to any individual, employee, former employee, or member of a business or government agency who reports misconduct to people or entities with the power and a presumed willingness to take corrective action. The misconduct is a violation of a law, rule, or regulation, or a direct threat to the public interest such as fraud or bribe taking. The UN and the development banks such the European Bank for Reconstruction and Development (EBRB) offer whistleblower protection.[7] An employee who complies in good faith with her or his duty to report suspected misconduct and who discloses information is protected from "pressure, retaliation or reprisal" in connection with the cooperation. Protection extends throughout the disciplinary and appeal process, including guarding the reputation of the whistleblower. Such policies could be introduced into any government organization or company engaged in dam planning and management.

Public participation and education tools.   Public education builds public trust. Those delivering this education are participants in the political, social, and bureaucratic arrangements that monitor and coordinate anticorruption strategies. A variety of public information programs can be launched in connection with a particular project or program. Specific tools include public meetings, media campaigns, school programs, public speaking engagements, and publications.

Ethics training.   Professional staff and employees at all levels would receive training on business ethics and ethical practices for government workers. Orientation training is a first step toward increasing employees' awareness, not only of the detrimental effects of corruption but also of the penalties.

Anticorruption radio spots and media campaigns.   These tools are part of public information and awareness programs.

Mapping Anticorruption Tools along the Project Cycle

Table 3.3 illustrates where the anticorruption tools just described are applied along both the Bank project cycle and the infrastructure project cycle, as discussed in chapter 2. Each tool has a communication component that is reflected in the overall project communication strategy.

Table 3.3. Mapping of Anticorruption Tools along the Project Cycle

	World Bank project cycle, stages					Infrastructure project cycle, stages				
	Macro planning (CAS)	Project identification	Project preparation, appraisal, approval	Project implementation	Project evaluation	Strategic planning	Project preparation and design	Construction/ implementation	Operation	Rehabilitation/uprating/ reoperation
National level										
National integrity system (NIS)	✓					✓				
National integrity pact (NIP)	✓					✓				
Access-to-information laws	✓					✓				
Anticorruption laws	✓					✓				
Lobbyist registration	✓					✓	✓			
Complaints and ombudsman office	✓		✓			✓	✓			
Independent anticorruption agencies	✓					✓				
Office for access to public information										
Sector level										
Business principles for countering bribery	✓					✓	✓			
Minimum standards	✓						✓	✓	✓	✓
Codes of ethics	✓						✓	✓	✓	✓
Public service disclosure of income and assets	✓			✓		✓	✓	✓	✓	
Oversight committees				✓						
Independent audit function				✓	✓			✓	✓	✓
Accountability benchmarks				✓	✓			✓	✓	✓
Survey and monitoring perceptions of corruption				✓	✓			✓	✓	✓
Project level										
Corruption risk assessments		✓	✓			✓	✓			✓
Governance improvements plans embedded in projects			✓	✓		✓	✓	✓	✓	✓
Integrity pacts for public procurement			✓	✓			✓	✓	✓	✓
Project anticorruption system (PACS)			✓	✓			✓	✓	✓	✓
Minimum standards			✓	✓			✓	✓	✓	✓
Codes of ethics			✓	✓			✓	✓	✓	✓
Project compliance plans			✓	✓			✓	✓	✓	✓
Other project-specific tools to incorporate in a GIP										
Participatory budgeting and budget tracking systems			✓					✓	✓	✓
Citizen Report Cards as an aid to public accountability in service provision and access				✓	✓			✓	✓	✓

continued

Table 3.3. *Continued*

	World Bank project cycle, stages					Infrastructure project cycle, stages				
	Macro planning (CAS)	Project identification	Project preparation, appraisal, approval	Project implementation	Project evaluation	Strategic planning	Project preparation and design	Construction/ implementation	Operation	Rehabilitation/uprating/ reoperation
Whistleblower protection	✓		✓	✓		✓	✓	✓	✓	✓
Public participation and education tools			✓			✓	✓	✓	✓	✓
Ethics training		✓	✓			✓	✓	✓	✓	✓
Anticorruption radio spots and media campaigns			✓			✓	✓	✓	✓	✓

Note: A communication strategy is prepared for each tool. Project advocacy argues for national and sector measures on Bank-supported projects to help demonstrate good practice. CAS = country assistance strategy.

Sustainability Tools

Figure 3.2, which also appears in chapter 1 of this handbook, depicts a framework of process- and performance-oriented themes that are relevant to delivering sustainability in dam projects. The various themes and their supporting tools and techniques reinforce and overlap.

Anyone viewing dams in the context of a wider development intervention will find that sustainability tools can help deliver the wide governance reforms needed to encompass dam planning and management and sustainable water resource management.

As emphasized in this handbook, the governance and anticorruption theme is an integral part of the sustainability equation. Thus the anticorruption tools discussed in the previous section are part of the "tool kit" that infrastructure practitioners need to deliver sustainability.

The sections that follow provide thumbnail sketches of a selection of existing and new tools to deliver each of these concepts. Each section begins by illustrating the relevance of each element to advancing the sustainable performance of dams and the value that communication adds in implementation. These different aspects of sustainability are intertwined in the development and management of infrastructure such as dams. It is not enough to look at each aspect in isolation—effective communication ensures key linkages.

The final section explains how these tools are mapped along the project cycle.

Integrated Water Resource Management (IWRM) Tools

The mission of the Bank–Netherlands Water Partnership Program (BNWPP) is to improve water security by promoting innovative approaches to integrated water

Figure 3.2. New and Old Sustainability Themes in Dam Planning and Management

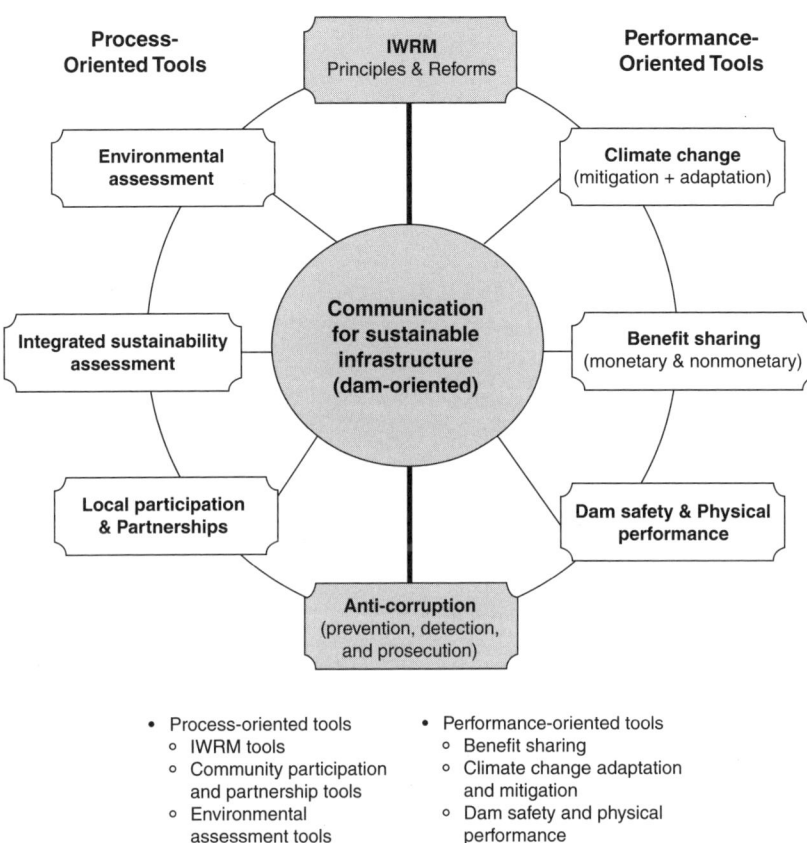

- Process-oriented tools
  - IWRM tools
  - Community participation and partnership tools
  - Environmental assessment tools
  - Integrated sustainability assessment tools
- Performance-oriented tools
  - Benefit sharing
  - Climate change adaptation and mitigation
  - Dam safety and physical performance

Source: Lawrence J. M. Haas.

resources management, thereby contributing to poverty reduction. The Dublin-Rio principles for IWRM place the planning and management of dams in a river basin context. In theory and in practice, IWRM reforms achieve the following:

- Reorient decision-making criteria for dam planning and management and involve a wider range of stakeholders in decisions about hydropower, including options on site selection, design, and operation (e.g., those with impacts on water allocation, quality, and use)
- Offer a wide variety of IWRM tools that are directly relevant to dam planning and management or at a minimum require linkage to IWRM tools
- Increase the need for communication and dialogue mechanisms among dam practitioners, river basin entities, and various basin-level, river use, and water use committees

What value can communication add?

- The establishment of clear priorities and issues on which to engage IWRM stakeholders

- A better understanding of the positions taken by the various IWRM stakeholders, of the project information they need, and of the methods of addressing their concerns
- Structured processes to engage IWRM stakeholders effectively in activities ranging from hydropower ranking studies to dialogue about the design and operating parameters of reservoirs and their impacts on competing water uses, as well as ecological and cultural values
- The deployment of conflict resolution tools where needed—for example, on water access and rights transformed by flow regulation and dam operations
- More informed public briefs and timelier access to information to create conditions for collaboration and avoid unnecessary conflicts caused by lack of accurate information
- An improved capacity to engage with and involve the media in linking the effects of dam operations to IWRM principles and water security concerns for all users

The Global Water Partnership (GWP) is perhaps the most recognized network promoting IWRM adoption at the national, regional, and global levels. The IWRM Toolbox maintained by the GWP provides a compendium of good practice instruments and 50 different tools to facilitate priority actions to improve water governance and management.[8]

The IWRM Toolbox has three categories of tools: (1) tools to create the enabling environment—that is, laws, policies, incentives; (2) tools to build appropriate institutions and capacity within institutions; and (3) management tools to deliver different aspects of IWRM principles.[9]

For illustration, the GWP's crosscutting management tools are listed in box 3.2. It is important to recognize the communication elements from advocacy to supporting partnerships. The BWP case study also illustrates the essential role of communication in the IWRM.

Community Participation and Partnership Tools

Most governments have explicit pro-poor policies that are part of their infrastructure development strategies linked to the Millennium Development Goals. Many countries have also instituted governance reforms to increase local participation in the decision-making processes in the water and energy sectors. Local participation is based on these tenets:

- Local community participation is central to the design and implementation of pro-poor measures in dam projects, responding to issues of voice, empowerment, and opportunity.
- Local action is often central to plan, implement, and monitor the social and environmental sustainability components of dam projects.
- Local action is needed to successfully introduce many anticorruption measures such as Citizen Report Cards to track anticorruption behavior.

Box 3.2. IWRM Management Tool Classification

C1. WATER RESOURCES ASSESSMENT—Understanding resources and needs
   C1.1 Water resources knowledge base
   C1.2 Water resources assessment
   C1.3 Modeling in IWRM
   C1.4 Developing water management indicators
   C1.5 Ecosystem assessment

C2. PLANS FOR IWRM—Combining development options, resource use, and human interaction
   C2.1 National integrated water resources management plans
   C2.2 Basin management plans
   C2.3 Groundwater management plans
   C2.4 Coastal zone management plans
   C2.5 Risk assessment and management
   C2.6 Environmental assessment (EA)
   C2.7 Social impact assessment (SIA)
   C2.8 Economic assessment

C3. EFFICIENCY IN WATER USE—Managing demand and supply
   C3.1 Improved efficiency of use
   C3.2 Recycling and reuse
   C3.3 Improved efficiency of water supply

C4. SOCIAL CHANGE INSTRUMENTS—Encouraging a water-oriented society
   C4.1 Education curricula on water management
   C4.2 Communication with stakeholders
   C4.3 Information and transparency for awareness raising

C5. CONFLICT RESOLUTION—Managing disputes, ensuring sharing of water
   C5.1 Conflict management
   C5.2 Shared vision planning
   C5.3 Consensus building

C6. REGULATORY INSTRUMENTS—Allocation and water use limits
   C6.1 Regulations for water quality
   C6.2 Regulations for water quantity
   C6.3 Regulations for water services
   C6.4 Land use planning controls and nature protection

C7. ECONOMIC INSTRUMENTS—Using value and prices for efficiency and equity
   C7.1 Pricing of water and water services
   C7.2 Pollution and environmental charges
   C7.3 Water markets and tradable permits
   C7.4 Subsidies and incentives

C8. INFORMATION EXCHANGE—Sharing knowledge for better water management
   C8.1 Information management systems
   C8.2 Sharing data for IWRM

Source: Global Water Partnership.

- Local action and community participation are needed to translate other governance reform themes to action on the ground (e.g., IWRM reform, dam safety, or payment for ecosystem services of value to sustainable management of reservoirs and dam-related assets).

What value can communication add?

- More effective integration of the overlapping aspects of community participation for different governance reforms (e.g., anticorruption, IWRM, local revenue sharing)
- Streamlined processes and reduced complexity in managing relations with local communities; clarity on accountability questions
- Improved climate of trust, good will, and cooperation to reach agreements with local communities adversely affected by hydropower
- Direct job creation and information on the new employment opportunities created by the project—empowerment + opportunity
- Exchange of ideas and experience on community-driven development initiatives linked to sustainable management of the hydropower asset, such as watershed management measures
- Communication mechanisms that empower the voice and choice of local communities and see that community needs are clearly understood and acted upon
- Help to build coalitions for local action to implement governance reforms.

Many NGOs and development organizations offer a range of participant tools to use in dam planning and management, especially those components of projects that directly involve local communities either contributing to decisions or relating to environmental and social mitigation and management.

Today, a variety of the approaches or frameworks for community participation incorporated into compensation and resettlement programs on dam projects are contextually defined. Increasingly, dam projects are including livelihood restoration programs (generally time-bounded) and actual long-term benefit-sharing mechanisms (benefit sharing is discussed later in this chapter). Apart from resettled people and resettlement host communities, involvement of the communities of traditional river users and land users in catchment is essential.

The World Bank's Participation and Civic Engagement Group focuses on the following themes for local community engagement in development projects—themes that encompass dam projects:

- *Social accountability* promotes the participation of citizens and communities in exacting accountability from the primary institutional actors in the government and the private sector.
- *Enabling environment for civic engagement* promotes conditions that enable civil society to engage effectively in development policies and projects over all stages of the project cycle.
- *Participatory monitoring and evaluation (PM&E)* promotes the participation of local beneficiaries in the monitoring and evaluation of projects and

programs, which is a key consideration in the sustainable performance of dams and governance and anticorruption.

- *Participation at the project, program, and policy levels* promotes participatory processes and stakeholder engagement at the three levels.[10]

Two illustrations of the supporting tools that help to indicate communication needs follow.

PM&E. A range of tools and methods are available for carrying out PM&E, but their relevance and applicability depend on the context. Anyone choosing PM&E approaches should ask several important questions. First, who leads and who follows? Externally led or project staff and representatives of beneficiaries co-design and manage the entire cycle (as depicted in the figure above). Second, what is the purpose of PM&E? The answer is project planning and understanding and negotiating stakeholder perspectives or public accountability.

Social accountability. These mechanisms are applicable within various contexts to build citizen voices and create spaces for the more proactive engagement of citizens and civil society with project authorities and with formal decision processes. These mechanisms include citizen advisory boards, vigilance committees, public interest litigation, public hearings, citizens' charters, and the right to information.

These communication tools and social accountability tools can be applied to budget questions, as noted in figure 3.3.

Figure 3.3. Communication and Social Accountability Tools Applied to Budget Questions

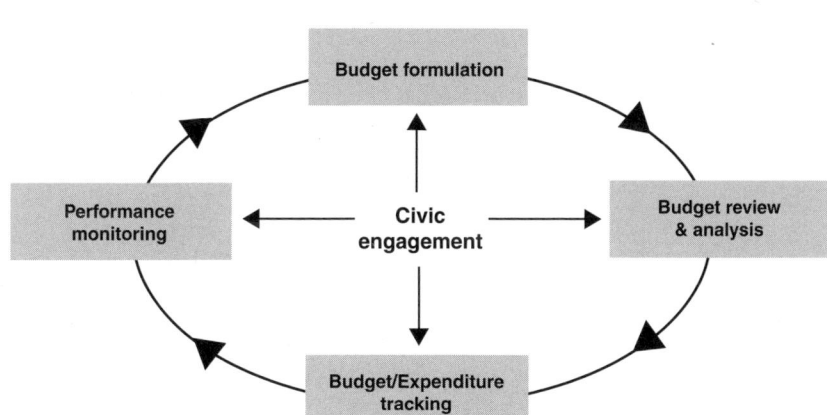

For dam projects, there is considerable potential in involving communities in social and environmental management budgets and also in local area development and revenue sharing. This involvement ensures that the social and environmental management components address the concerns that are relevant to local stakeholders and improve their perceptions of the social and environmental performance of the project. At the same time, these measures improve community relations and enhance the overall effectiveness of the dam as a development intervention.

Environmental Assessment Tools

A broad range of environmental assessment and management tools span the family of strategic environmental assessments (SEAs) and cumulative impact assessments (CIAs) used in up-front planning of the operation stage environmental monitoring and management plans (EMMPs) and comprehensive environmental management systems.[11] Today, communication is generally recognized to be an essential part of environmental assessment and management tools. Effective communication is essential to identifying the environmental performance risk faced by all stakeholders, from the livelihood risks faced by local communities to the reputational risks faced by financing partners. These tools inform and invite participation:

- Dam projects have and continue to be criticized for real or perceived lack of information access, public consultation, and meaningful engagement with affected groups on various types of environmental assessments from SEAs to EMMPs.
- New tools such as environmental flow assessments (EFAs) can substantially contribute to the environmental performance of hydropower projects, but few people are aware of the value of these tools, understanding of them remains low, and they are confused with older practices.
- The scope for advocacy to implement the current generation of good practice environmental assessments is considerable.
- Most environmental assessments have substantial explicit and implicit communication needs. Failure to provide effective two-way communication undermines their value and credibility, which leads to controversy over dam projects or a suboptimal design, thereby increasing the risk exposure for all stakeholders.

What value can communication add?

- Improving the design and management of communication components for environmental assessments assisting with project approval and financial support.
- Encouraging private sector lenders to undertake more effective, reliable, and accurate due diligence of the environmental components of dam projects.

- With more effective stakeholder engagement, identifying environmental concerns early, focusing efforts on the problems, improving the quality and acceptance of environment assessments, and reducing the risk of unnecessary controversy.
- Capturing and reflecting the local knowledge and best available information in environmental assessments and mitigation and monitoring programs.
- Overcoming the technical language and jargon barriers that stand in the way of local understanding of and local inputs to assessments.
- Applying relevant lessons from other projects in the country or region.
- Conveying an understanding of environmental stewardship and the respective roles of the project authorities, government agencies, and local communities.

A range of well-known but constantly evolving environmental assessment tools includes those for strategic and project-level work. At the sector, basin, or strategic level, they are strategic environmental assessments, cumulative impact assessments, and environmental flow assessments. At the dam project level, they are project-level initial environmental examinations (IEEs), environmental impact assessments (EIAs), and so on; environmental monitoring and management plans; and environmental flow assessments.

A considerable body of literature describing these tools has been under development since the 1970s and 1980s. The CIA and EFA tools are the more recent in terms of refinement of techniques and their specific application to promoting sustainability in dam planning and management. The following description of EFA tools highlights their relevance at different stages of the project cycle and as communication-intensive in nature.

Environmental flow assessment (EFA) and provision.   An environmental flow (or ecological flow in some jurisdictions) is the provision of water within rivers and associated ground water systems in sufficient quality, quantity, timing, and duration to maintain freshwater ecosystems and wetlands and their benefits where the river is subject to competing uses and flow regulation. Various international organizations have compiled tools for environmental flow assessment and provision, including organizations such as IUCN, World Wide Fund for Nature, World Bank, and the GWP.

Figure 3.4 illustrates three general categories of EFA tools and applications:

1. *Policy and regulatory tools.* Policies and regulations on river flows, ranging from prescriptive to negotiated outcomes.
2. *Quantification tools.* Four main families of methods in order of increasing complexity and data intensiveness:
   - Index methods/look-up tables
   - Rating methods/desktop analysis
   - Functional methods/expert panels
   - Physical habitat models

Figure 3.4. Environmental Flow Assessments in Integrated Water Resource Management

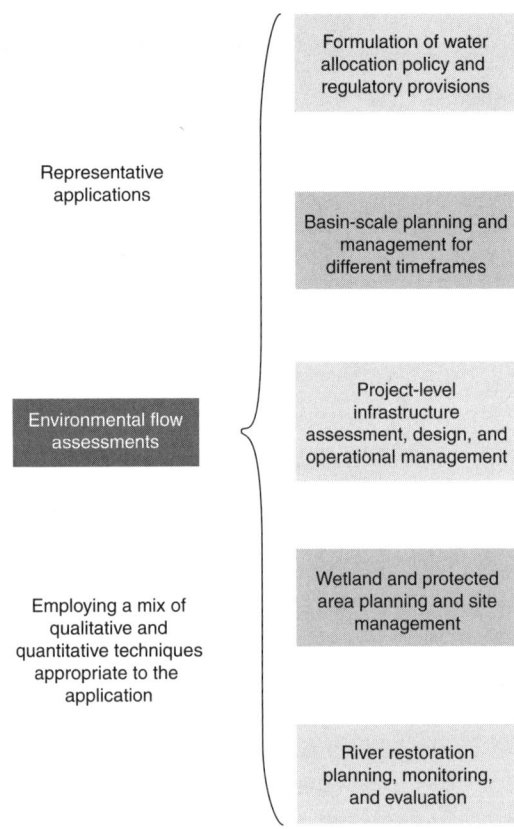

Source: International Union for Conservation of Nature, Mediterranean Flows Pamphlet, http://iucn.org/places/medoffice/en/en_wetland.html.

3. *Broader frameworks.* More in-depth procedures, incorporating one or more of the quantification tools into wider assessment frameworks:
   - Downstream Response to Imposed Flow Transformation (DRIFT)
   - Instream Flow Incremental Methodology (IFIM)

The IFIM framework, which was developed in the 1970s in the United States, has undergone considerable transformation over the years. Legally mandated in a number of jurisdictions, it is an integral part of hydropower licensing procedures. The DRIFT framework, developed in South Africa, addresses all aspects of the river ecosystem. It employs four modules to construct scenarios and their ecological, social, and economic implications, which are developed in a collaborative process with river users and community stakeholders.

The DRIFT methodology is featured in the two case studies in appendixes A and B. In fact, both of these projects are pioneering efforts to implement environmental flow assessments, starting with pre-project baseline assessments and then linking monitoring to adaptive management of the reservoirs. In the BWP project, environmental flows were essential for river ecology and maintaining the critical functionality of ecosystems, and the environmental flow releases were critical for maintaining water quality in downstream river reaches to support a variety of water use needs, including irrigation for high-value crops that are the mainstay of the local economy and offtakes for the municipal water supply, tourism, and recreational values.

Communication platforms and dialogue mechanisms were critical at all stages of approval of the Berg Water Project and are maintained during the operational phase.

Integrated Sustainability Assessment Tools

Tools to identify sustainability in dam projects and to screen out controversial projects from the national inventory of potential sites at an early stage are continually evolving. Such assessment tools are now used as indicators of eligibility for financial support (e.g., requirements by the Organisation for Economic Co-operation and Development to support sustainable hydropower projects)[12] or as rapid appraisal techniques to identify aspects of hydropower and multipurpose projects that require additional scrutiny or due diligence. Another dimension of sustainability is the financial sustainability of service provision and financing to meet ongoing environmental and social management commitments, linking to tariff reform policies. The present sustainable guidelines include the following:

- Low-impact hydropower certification schemes are now available or proposed in North America and Europe.
- The International Hydropower Association (IHA) has developed voluntary protocol and sustainability guidelines endorsed by industry and IHA members.
- Multistakeholder initiatives are under way to widen the acceptance and utility of the current sustainability guidelines such as those of the IHA.
- Mechanisms, such as using a portion of project revenue to ensure the financial sustainability of environmental and social management components of hydropower projects, exist, but they are not used widely (e.g., to fund catchment management also linked to revenue sharing).

Communication will encourage public acceptance of tariff levels that internalize the social and environmental costs of hydropower projects, leading to the financial sustainability of reliable levels of service and of the social and environmental management components of hydropower projects.

What value can communication add?

- An advocacy strategy to encourage the adoption of sustainability criteria and guidelines in hydropower ranking studies to screen out unsustainable projects early
- Tools to evaluate what aspects of design need to be improved to enhance their sustainability

- A clear message to the public on why it is important to internalize the environmental and social costs of hydropower projects and generation options in tariffs and processes, leading to a dialogue on the implications for sustainable tariffs
- Voluntary sustainability assessments to identify problem areas in independent power producers (IPPs) or privately financed projects in settings where environmental standards may be weak or the institutional capacity to deliver national standards is low
- A consensus on what criteria apply to assessing the sustainability credentials of proposed hydropower projects
- Better feedback on what electricity consumers will accept or trade off in the selection of generation options and the effect on retail tariffs.

Advice on how good sustainability scores can enhance the financing prospects for hydropower projects and access to innovative financing such as carbon financing.

This sustainability theme has three dimensions: (1) sustainability protocols and guidelines (for hydropower and dam projects generally); (2) financial sustainability of operations for service provision; and (3) physical performance sustainability of hydropower assets and dam infrastructure.

Sustainability Protocols and Guidelines

The International Hydropower Association took the industry lead in developing protocols and guidelines for hydropower. Other low-impact certification tools are available. Some existing and potential future applications of protocols and guidelines are to score projects against the guidelines to evaluate a portfolio of projects, identify areas requiring greater attention or due diligence, establish the eligibility of projects for loan financing for different agencies, evaluate eligibility for carbon financing, and certify projects for green financing labeling. Each of these applications needs communication support.

Sustainable Operations

The financial sustainability of existing and planned hydropower is important from all perspectives—that is, those of the operators, government, and electricity consumers. Two aspects are critical: (1) financially viable operations and (2) sustainable financing of environmental and social management components to meet commitments.

The tools are generally part of the wider set of financial and economic evaluation tools and tariff evaluation tools. From a communication perspective, a central issue is to provide clear and consistent messages to stakeholders and electricity consumers about the need for tariffs to enable financial sustainability.

Physical Sustainability

Proper maintenance is central to ensuring the smooth performance of existing structures and equipment on a day-to-day basis. However, it cannot fully eliminate the aging process to which all large infrastructure projects (such as bridges, roads, buildings, ports, and dams) are subjected. It is therefore necessary to occasionally undertake rehabilitation projects in order to avoid both lower revenues and higher operation and maintenance costs.[13]

Figure 3.5 illustrates how the performance established at the commissioning of a project continues for a number of years until it begins to deteriorate. The rate and level of the duration depend on project-specific circumstances. In certain types of projects in which, for example, live storage is lost over time, the project may become uneconomic. If the project is not maintained, failure is a risk (see section on dam safety tools later in this chapter).

Otherwise, restoration works can maintain the project at or near its original performance levels. Depending on the situation, hydropower projects can also be uprated. Beyond that, there is room for nonstructural optimization to improve performance, such as how the hydropower project is operated in the grid system. As the WCD has reported, nonstructural optimization can increase performance by 15 percent or more, but the opportunities are case-specific.

With physical sustainability, communication again comes into the picture: (1) it involves stakeholders in decision processes in the same way they

Figure 3.5. Performance at Commissioning

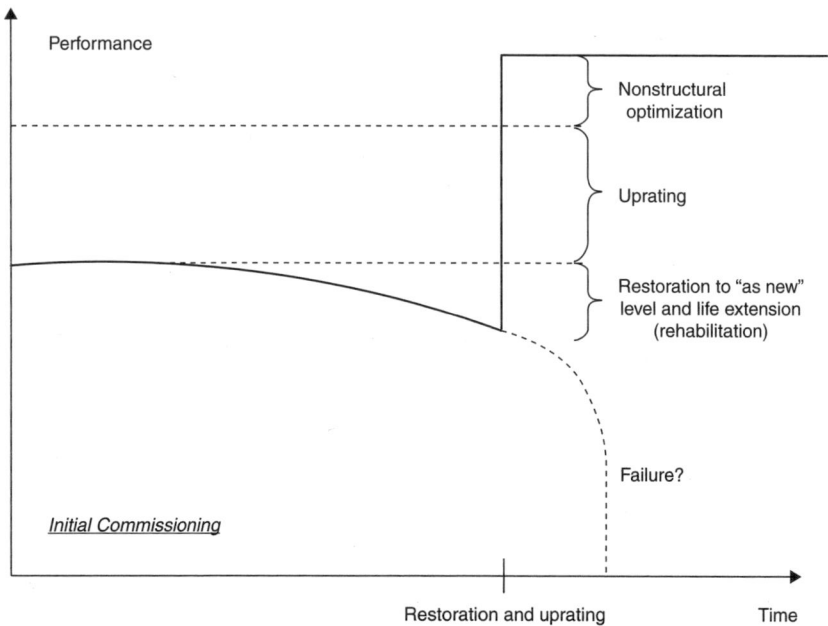

Source: Lawrence J. M. Haas.

participated in the decisions about the selection, design, and implementation of the original project; (2) it offers opportunities to enhance the social and environmental performance at the same time as the technical aspects; and (3) it reflects the other governance themes discussed that contribute to sustainable hydropower development.

Benefit-Sharing Tools

Benefit sharing is increasingly attracting attention worldwide as a uniquely powerful, practical, and adaptable management tool. It underpins the kinds of partnerships needed to genuinely involve people in the development decisions that affect them and to put IWRM principles for the sustainable management of large dams into practice. Benefit sharing occurs at different levels, from sharing between riparian states to sharing national benefits with basin and dam-affected populations and river users. Some considerations follow:

- Many observers argue that benefit sharing is a necessary condition for multi-country cooperation on international rivers. By focusing on the division of benefits that derive from water use, not the physical allocation of water, planners can identify substantive mutual gains.[14]
- Local benefit sharing has three forms: (1) equitable sharing of infrastructure services; (2) nonmonetary benefits such as enhanced access to natural resources to replace the loss or transformation of access before the project; and (3) monetary forms of revenue sharing with the local community or the wider basin community that hosts the project.
- Many countries now have legislation in which local communities that give up their land entitlements or access to natural resources receive a monetary share of the project revenues, but revenue sharing is not yet common practice. In some countries, revenue sharing funds local action for sustainable management of the dam assets such as tree planting in headwater forests or payment for ecological services, which have mutual benefits for both the local communities and dam operations.[15]
- The design and implementation of benefit-sharing measures have greater and different communication needs—indeed, much greater than those for compensation and resettlement stages—and require long-term communication capacity building. Rampant corruption or abuse of power in the administration of revenue sharing can seriously undermine public confidence.

What value can communication add?

- Offering structured processes for dialogue with riparian states or local communities on the design and set-up of benefit-sharing measures
- Linking the dialogue on benefit sharing with the community dialogue on the social and environmental management components of the dam project,

giving new dimension and meaning to partnership philosophies and meeting expectations

- Integrating benefit sharing into the regular local development processes and ensuring that the government will not reduce the regular local development entitlements of those people anticipating benefit-sharing programs
- Developing a clear sense of corruption risks and vulnerabilities and a clear sense of social accountability needs and achieving a consensus on how to tackle the concerns raised and effective action plans
- Advocating more effectively at the sector or regulatory levels to introduce benefit-sharing policies and improve understanding on how they apply to hydropower projects
- Enabling the use of community-driven development (CDD) approaches and the use of revenue-sharing funds according to beneficiary preferences (bottom-up and not top-down)
- Enabling local communities to exchange information, ideas, and approaches to maximize the benefit of revenue-sharing funds
- Establishing mechanisms such as trusts and setting up representative boards for governance to ensure full local ownership and local management responsibility

Benefit sharing has recently come to the forefront of international thinking about ways to develop and manage large dams sustainably and distribute the benefits and costs more equitably within society.[16] Sharing benefits equitably is a way to deliver value for all stakeholders, as well as a practical approach to catalyzing and funding local actions that join together many strands of water governance reform and sustainable thinking within the IWRM framework. Benefit-sharing mechanisms reinforce social equity in infrastructure strategies and promote sustainability rather than narrowly optimize dams as physical assets that deliver water and energy services.

Many countries now have policies under which local communities that give up their land entitlements or access to natural resources receive a monetary share of the project revenues, but examples are still few and far between. In Vietnam, sustainable development is a central theme of the 2004 Electricity Law. It is defined as "development that meets the needs of the present generation without compromising the ability of future generations to meet their own needs, on the basis of a close and harmonized combination of economic growth, assurance of social advancement and environmental protection."[17]

The general principles of benefit sharing apply equally to sharing between riparian states and sharing between the national and local levels.[18] Three broader approaches can be taken to sharing the national benefits of dam developments with local communities and river basin populations. First, in *equitable sharing of project services,* local populations, as target beneficiaries, receive equitable access to the water and energy services produced by dam projects to support their development and welfare opportunities.[19] Second, in *nonmonetary forms of benefit sharing,* the target beneficiaries receive entitlements giving them access to other natural resources or support to pursue other forms of livelihood and welfare that offset the permanent loss of or reduction in access to land or water resources caused by the dam.[20] And, third, in *revenue sharing,* targeted beneficiaries share part of the monetary benefits generated by the project, typically in the form of a portion of the revenue from bulk electricity sales or bulk water sales on an annual basis.

These arrangements are generally maintained over the economic life of the dam project. They commence after the project becomes operational. Other forms of benefit sharing may begin during the project implementation stages, which can span several years. These forms include investments to maximize local employment in the construction workforce and local supply of goods and services to the project, as well as investments in physical infrastructure such as local roads (e.g., those that increase a community's accessibility to agricultural markets or, for villages near reservoirs, to health care) and other public services that have sustainable long-term benefits for communities.

The notion of local benefit sharing goes beyond a onetime compensation payment and short-term resettlement support for displaced people. It treats both displaced people and the communities that host the hydropower project and give up land or access to other resources as legitimate partners in the project and first among its beneficiaries.

The supporting tools include advocacy tools for legislation that enables the establishment of benefit sharing, community participation tools, social accountability tools, and tools that support community-driven development approaches. These tools are inherently participatory and communication-intensive in nature.

Climate Mitigation and Adaptation Tools

Climate change has many implications for the provision of water and energy services and for hydropower development from two perspectives: that of mitigation (i.e., reducing the emission of greenhouse gases in electricity generation) and that of adaptation (i.e., adapting dam delays and other hydropower structures and water resource systems to increasing climate variability and extremes). Adaptation and mitigation raise the following needs:

- On the mitigation side of the equation, hydropower offsets thermal greenhouse gas (GHG) emissions, but the potential for reservoir emissions should be taken into account and subtracted. On the adaptation side of the equation, drought and flood extremes need to be factored into dam planning, design, and management in transparent and explicit ways.
- The combined effects of climate variability in terms of changes to the hydrological cycle and the loss of the hydrological contribution to ecosystems transformed by dam operations need to be better understood and the best available information applied.

- Opportunities to access international financing sources for carbon financing are available for hydropower. However, these tools are still nascent and have high transaction costs. Controversy over the eligibility of hydropower remains, along with a lack of clarity, particularly on how storage projects are treated.

Operational tools also need to be revamped to address, for example, impacts on the "fill vs. spill" dilemma and planning for energy reliability (in times of drought) and to introduce sediment management strategies to cope with accelerated soil erosion and reservoir sedimentation.

What value can communication add?

- Better understanding of the questions surrounding climate change adaptation and mitigation relevant to the project and a consensus on what actions to reflect in it
- Better local participation in climate vulnerability and the risk assessments needed to factor into strategic planning and project design activities
- Clearly structured processes for an informed dialogue with national and international stakeholders on the climate issues that arise on the project
- Ability to mobilize efforts to deal with any problems that may arise with the application for or use of carbon financing tools to enhance the sustainability of the project (e.g., Clean Development Mechanism eligibility, registration, or selling carbon credits to third parties)
- Tapping previous experience and in-country expertise to exchange information and to understand the opportunities, complexities, and limitations of carbon trading
- Better links between the national scientific community involved in forecasts of climate variability and extremes and practitioners involved in dam planning and management
- Ability to inform national and local stakeholders of progress in securing carbon financing support and decisions on the use of funds
- Strategy to inform the public and special interest groups on how projected climate variability and extremes are taken into account in the project identification stage and in the design and operating strategies of the project.

A variety of tools are being developed to address the risks to the sustainable performance of dams arising from climate variability. Many tools for climate mitigation and adaptation are directly relevant to the water sector.[21] Some specific tools are appropriate for hydropower, focusing on drought and flood extremes, including vulnerabilities from a planning and design perspective, from an operational perspective, or from an environmental or water regime perspective.[22]

Proactive and planned climate change adaptation is a relatively new challenge in dam planning and management. The need to retool practices and adjust design criteria is generally recognized by dam practitioners—for example, the impact of the probable maximum flood (PMF) on the design and capacity of spillway structures and the storage capacities needed to respond to changes in long-term flow, seasonal variability of flow, and seasonal changes in demand.

Operational tools also need to be revamped to address, for example, the "fill vs. spill" dilemma, to plan for energy reliability (such as in times of drought), and to introduce sediment management strategies to cope with increased reservoir sedimentation.

A variety of communication tools are needed to link actions to wider river basin management reforms and processes. And water users should receive help in "climate proofing" ecological and dam services. For example, drought management policies could be introduced that take into account extractions from reservoirs, water courses, and downstream reservoir releases in order to maintain the critical functionality of wetlands or other vulnerable areas during deeper and more prolonged periods of drought.

The following approaches and tools could be used as well:

- Information providers to help in risk assessment:
  - For spatial-temporal hydroclimatic state variables: NAPA Platform, PRECIS[23]
  - To assess hydrological impacts: computer-based models such as DGIS or ORCHID project design tools
- Screening tools and processes to support local adaptation to climate change:
  - Integrating climate change considerations and adaptation into current risk management strategies and planning processes such as impact and vulnerability assessments at varying levels and scale
  - Using tools to assess the vulnerabilities of the social and environmental components of dam projects to climate change—for example, CRiSTAL[24]
  - Using guidebooks on adapting to climate changes such as those by the United Nations Development Programme (UNDP) and the U.S. Agency for International Development (USAID)

Climate Change Mitigation

The potential role of hydropower in reducing GHG emissions from the power sector is well recognized. Methodologies and tools are now available to help quantify the relative value of dams in mitigating GHG emissions and to facilitate maximizing the contribution. They include:

- Electricity generation portfolios and minimum portfolio standards in strategic planning (linked to emission reduction targets)
- Carbon trading, including the United Nations Framework Convention on Climate Change (UNFCCC)-sponsored Clean Development Mechanism (CDM) for trading certified emission reductions (CERs) measured

in terms of CO_2-equivalent verified emission reductions (VERs), which do not fall under the CDM

- Baseline carbon flux assessments (as part of project environmental impact assessments)
- Reservoir emissions mitigation and monitoring (various computer and field measurement techniques)
- Methodologies for assessing the carbon offset and the potential of alternative generation technologies in simple and more complex power systems (e.g., as evolving in the CDM process)[25]

Communication

Particularly in water-stressed basins, it is important that an open communication process is established with stakeholders to link the dam management strategy and associated catchment management measures to the risks facing water users, consumers of dam services, and basin communities associated with increased climate variability and extremes. The BWP case study described in appendix A notes that South Africa has taken the position that the IWRM framework captures many measures that would help river basins and dams adapt to climate change.

Climate change is still a relatively new field. In order to move forward, coping with climate change and creating local capacity, it is important to communicate these issues in the partnerships between the scientific community and the users of the information, and also within the dam industry itself.

Dam Safety Tools

Dam safety has received considerable attention in the past few decades. New regulatory frameworks and standards for dam safety have been or are being introduced or reinforced in most countries, especially in transitional economies. Dam safety issues and remedial measures are linked to other sustainability and governance reform themes (e.g., IWRM, anticorruption, sustainable finance, and climate change). Various factors such as the aging population of dams, the highly variable design standards of old dams, and the increased variability in the climate system (flood risk and higher PMF estimates) have brought attention to dam safety issues. Some observations about dam safety programs follow:

- Many of the tools for dam safety programs require information exchanges, technical cooperation, and coordination among dam owners, regulators, and dam safety officials.
- Dam safety programs also incorporate agency and public response preparedness measures.

- Although dam safety is important in new dams, the bulk of the work at present is in dealing with the existing dams.
- Dam safety programs have a variety of communication support needs for dialogue and information sharing among professional agencies and with local communities.

What value can communication add?

- Effective engagement of technical bodies and regulatory bodies in dam safety programs relevant to the project
- Better understanding of risks and how local communities and agencies responsible for emergency preparedness perceive them
- Structured processes for an informed public dialogue on dam safety issues
- Consensus on priority investments in dam safety for existing dams using inclusive and participatory processes; information packages to inform the public about the dam safety measures adopted
- Effective engagement with the local communities on emergency preparedness programs and awareness of response measures in the event of an emergency event
- Dialogue mechanisms that maintain local and public confidence in the safety and integrity of the dam structures and emergency preparedness.

Typically, dam safety programs utilize some tools that need varying degrees of communication support. One example of dam safety tools is Dam Safety Program Management Tools (DSPMT), an information collection and management system used by U.S. federal and state dam safety program managers to provide as-requested and periodic information on local dam safety, program needs, and accomplishments within each organization's jurisdiction.[26]

The DSPMT is a set of interactive software programs that can serve as a resource to dam safety data owners, managers, and data providers. It currently includes four distinct functional areas: (1) Dam Safety Program Performance Measures (DSPPM);[27] (2) NID Electronic Submittal Workflow; (3) Palm-based Inspection Checklists; and (4) the DSPMT Web site (http://www.safedams.org).

The tools applied on an individual project will depend on the outcome of the initial dam safety risk assessment.

The project-specific communication support needs can be inferred from the type of training offered in a comprehensive dam safety program:

- Technical training
  - Conducting continuing education courses for dam safety engineers and technicians

- Collecting and posting information about technical training available from other organizations and agencies
  - Sponsoring state personnel to attend continuing education
- Support for state programs
  - Disseminating public and dam owner awareness tools for state dam safety programs
  - Compiling guidelines for dam safety programs
  - Offering peer review services to help states recognize their weaknesses and improve their programs
- Partnering
  - Strengthening the cause by bringing together organizations and agencies involved with dam safety
  - Working jointly with other organizations on projects to improve dam safety
- Public awareness
  - Sponsoring public awareness workshops
  - Educating state and federal lawmakers about the need for strong dam safety programs
  - Working with the media to spread the message about the importance of dam safety to the public
- Information exchange
  - Maintaining a clearinghouse of information on dam safety
  - Reaching the public through the Internet
  - Providing research services
  - Producing educational publications and periodicals and statistical analyses
- Networking
  - Offering avenues for communication between dam safety professionals
  - Promoting and organizing regional and state meetings to bring together dam owners and regulators
  - Bringing together people to present a strong message on dam safety to lawmakers and to the public.

Sustainability Tools along the Project Cycle

Table 3.4 illustrates where the sustainability tools discussed previously are applied along both the Bank project cycle and the infrastructure project cycle, as discussed in chapter 2 of this handbook.

Table 3.4. Mapping of Sustainability Tools along the Project Cycle

		Stage of project cycle at which tools are applied				
3, some / 2, more / 1, most		Macro level	Sector strategic Planning/ regulation	Project preparation	Construction phase	Operation phase
1.	**IWRM tools**					
	Improving the enabling environment					
	Creating the institutional environment					
	Management tools					
2.	**Community participation tools**					
	Social accountability					
	Enabling environment for civic engagement					
	Participatory monitoring and evaluation					
	Participation at the project, program, and policy levels					
3.	**Environmental assessment tools**					
	SEAs (sector, basin, etc.)					
	IEE, EIA, etc.					
	Environment management, mitigation, and monitoring					
	Environmental flow assessment					
4.	**Hydropower sustainability tools**					
	Sustainability protocol					
	Sustainability guidelines					
5.	**Climate mitigation and adaptation tools**					
	Emission reduction tools					
	Adapting dams					
6.	**Benefit-sharing tools**					
	Revenue sharing					
	Enhancing resource access entitlements					
	Equitably sharing project outputs					
7.	**Dam safety tools**					
	Dam safety regulation					
	Dam safety programs					
	Project-specific tools					

Notes

1. Institutional communications can encompass internal communications.
2. Transparency International, http://www.transparency.org/tools. TI has successfully applied some of these tools to hydropower projects and water infrastructure services in various countries in partnership with government agencies, developers, utilities, and civil society organizations.
3. The *Transparency International Country Study Report of Japan* is part of a 2006 series of national integrity system country studies of East and Southeast Asia, http://www.transparency.org/content/download/14728/156321/file/japan_r.pdf.
4. World Bank, http://go.worldbank.org/QXV28CEF90.
5. On April 17, 2006, Oil and Natural Gas Corporation became the first Indian company to sign an integrity pact with Transparency International. The pact was signed by company chairman and managing director Subir Raha and Admiral R. K. Tahliani, former chief of naval staff and chairman of TI India. The company had used the pact since July 2005. Developed by TI, the integrity pact is a tool to prevent corruption in public contracting. Transparency International, *TI Watch*, May 2006.
6. For details on this system, see Anti Corruption Commission, Republic of Sierra Leone, http://www.anticorruption.sl/addressing_corruption.htm.
7. See http://www.whistleblower.org/template/index.cfm and http://www.ebrd.com/about/strategy/general/whistle.htm.
8. Global Water Partnership, "ToolBox: Integrated Water Resource Management," http://www.gwptoolbox.org.
9. Ibid.
10. World Bank, Participation and Civil Engagement, "Tools and Methods," http://web.worldbank.org/WBSITE/EXTERNAL/TOPICS/EXTSOCIALDEVELOPMENT/EXTPCENG/0,,contentMDK:20509330~menuPK:1278210~pagePK:148956~piPK:216618~theSitePK:410306,00.html.
11. Examples are SEAs at the basin or sector level, project environmental impact assessments (EIAs), environmental flow assessments (EFAs), EMMPs, and CIAs.
12. See the 2003 Recommendation on Common Approaches on Environment and Officially Supported Export Credits [C(2003)236, as amended by Council in C(2004)213] and updated in 2005, http://webdomino1.oecd.org/olis/2005doc.nsf/43bb6130e5e86e5fc12569fa005d004c/db22f47098127953c1256fb3005ec629/$FILE/JT00179362.PDF.
13. Laurent Bellet, Jean-Michel Devernay, Donal O'Leary, and Alessandro Palmieri, "Framework for Policy and Decision Making on Dam and Hydro Plant Rehabilitation and Uprating," http://www.un.org/esa/sustdev/sdissues/energy/op/hydro_bellet_paper.pdf.
14. T. Yu and H. Winston, "Benefit Sharing in International Rivers: Findings from the Senegal River Basin, the Columbia River Basin, and the Lesotho Highlands Water Project," Report No. 46456, Africa Region Water Resources Unit Working Paper 1, World Bank, Washington, DC, 2008, http://go.worldbank.org/U87VJHWH20.
15. Examples of these benefits are capturing cross-sectoral synergies in land management, local income generation, and sustainable management of dams as physical assets; extending the operating lives of reservoirs by planting trees in headwater areas; or shifting to agricultural and livestock grazing practices that combat desertification, soil erosion, and sediment processes in river catchments—multiple benefits.
16. Beyond the dam sector, benefit sharing is actively pursued today in other natural resource extraction and transformation sectors. The numerous models from the mining, petroleum, and forestry sectors range from nationally administered revenue funds that target improvements in public services to affected communities, to

revenue-sharing contracts between companies (or state production enterprises) and local communities.

17. Article 4 of the Electricity Law (2004) calls for developing a "sustainable power sector based on optimal development of all sources." Article 1 of the revised Law on Environmental Protection (2005) that came into effect in July 2006 defines sustainable development for all sectors of the economy.

18. Multicountry arrangements are typically more complicated because of their cost-sharing dimensions, varying benefits for each country, and the decades-long agreements that can ensue if political relations between states are complex.

19. For electricity services, a range of measures can be considered such as mandatory electrification of resettlement communities; priority in rural electrification programs for connection or improved levels of service; financial assistance for individual household service connections, and possibly energy-efficient appliances such as lighting; or preferential electricity tariffs for a stipulated period of time.

20. Nonmonetary benefits can be as valuable to local communities as the monetary benefits, especially measures that empower and build local capacity for management of natural resources and access to ecosystem services. But they may also have an indirect cost. The cost may be minor, such as deferment of potential local tax revenue when local fishermen are granted preferential licenses for reservoir fisheries, or they may have a more measurable impact on overall project economics, such as when water is released from reservoirs to maintain recession agriculture downstream (though the net development and sustainability gain remains positive).

21. United Nations Framework Convention on Climate Change, http://unfccc.int/files/adaptation/methodologies_for/vulnerability_and_adaptation/application/pdf/water_sector_tools.pdf.

22. Canadian Climate Impacts and Adaptation Research Network, "Hydropower and Climate Change: A Workshop Report," 2006, http://www.c-ciarn.ca/pdf/hpccwksreport.pdf.

23. For more details, see United Nations Framework Convention on Climate Change, http://unfccc.int/adaptation/methodologies_for/vulnerability_and_adaptation/items/2674.php.

24. Project-based tools are piloted for undergoing pilot studies in Nicaragua, Mali, Tanzania, and Sri Lanka in sectors such as water resource management, infrastructure, and natural resource management. The tools deliver vulnerability and livelihood profiles, as well as details for project modification. For more information, see International Institute for Sustainable Development, http://www.iisd.org/security/es/resilience/climate_phase2.asp.

25. United Nations Framework Convention on Climate Change, http://cdm.unfccc.int/Panels/index.html.

26. For more information, go to http://www.safedams.org/ and http://www.fema.gov/plan/prevent/damfailure/information.shtm.

27. The DSPPM is made up of the following modules: DSPPM 1 - Dam Safety Program Management Authorities and Practices; DSPPM 2 - Dam Safety Staff Size and Relevant Experience; DSPPM 3 - Inspections and Evaluations; DSPPM 4 - Identification and Remediation of Deficient Dams; DSPPM 5 - Project Response Preparedness; DSPPM 6 - Agency and Public Response Preparedness; DSPPM 7 - Unscheduled Dam Safety Program Actions.

Appendix A
Berg Water Project

The Berg Water Project (BWP), which consists of the Berg River Dam and the supplemental diversion scheme, is designed to augment the raw bulk water supply to the greater Cape Town metropolitan area in the Western Cape Region of South Africa. The BWP captures and stores winter runoff from the mountainous upper reaches of the Berg River basin. The runoff is then transferred to the existing to Western Cape Water Supply System (WCWSS) to help meet 18 percent of the urban and irrigation demand during the dry summer periods. The river diversion project must also release enough water into the downstream of the Berg River system for a statutory "reserve" flow to meet the basic human and ecological needs and seasonal water needs of a variety of downstream river users and for water quality management.

Looking through the lenses of governance, sustainability, and communication, the BWP is an interesting case study in several respects:

- The BWP was the first major bulk water transfer scheme approved in the post-1994 era under South Africa's progressive water legislation, and was the largest water project in the country at the time (see box A.1). The design parameters and the decision to build the BWP were the first major test of the new legislation.
- The BWP richly illustrates a contextual translation of progressive macro policy reforms first into an infrastructure strategy and then into a project—and the challenges this translation entails.

Appendix A is adapted from Lawrence J.M. Haas, Leonardo Mazzei, Donal T. O'Leary, and Nigel Rossouw, *Berg Water Project: Communications Practices for Governance and Sustainability Improvement,* World Bank Working Paper No. 199 (Washington, DC: World Bank, 2010).

Box A.1. Berg Water Project: Its Genesis

The BWP was part of a multitrack drive to improve water security for over 3 million people served by the integrated Western Cape Water Supply System (WCWSS), where combined demand from urban and agriculture users will exceed the water yield available from conventional water resources in the area well before 2020.

The initial project planning occurred during fundamental governance transformations in South Africa in the 1990s. The project was eventually approved in 2002, but only after exercises in cooperative governance involving interested and affected parties (I&APs) in three parallel public processes (1) on assigning priority to water demand and supply reconciliation options in the Berg Water Management Area, (2) on the project environmental impact assessment (EIA), and (3) on Cape Town's targets and investment plan for water services improvement, a statutory requirement.

Cape Town residents were subject to stringent curbs in water use during the drought of 1998–2000 when these decisions were in process. This situation had an effect on public perceptions about water security and deeply polarized views about how to deal with scarcity. Many environmental nongovernmental organizations and right-based civil society organizations were firmly against the BWP, contending that demand management was a better investment and the only environmentally sustainable solution. The City of Cape Town argued that demand and supply measures were complementary and that both were urgently needed.

The BWP was approved by the South African Cabinet conditional on Cape Town taking steps to ensure a 20 percent reduction in projected water demand by 2010. The Trans-Caledon Tunnel Authority (TCTA), a state-owned entity mandated to implement raw bulk water infrastructure in South Africa, began construction in 2004, and the project became operational in 2008.

- As the formative Catchment Management Authority (CMA) and its supporting water user entities gradually become functional, the BWP promises further insights on adapting the operation of a dam to a catchment management strategy founded on integrated water resource management (IWRM) principles and competing ecological, economic, and social values.

Macro Planning and Project Preparation Stages

In 1966 South Africa's new constitution introduced the notion of cooperative governance to involve people in the development decisions that affect them. The constitution's principles were translated into South Africa's much acclaimed water laws by the Department of Water Affairs and Forestry (DWAF) in open processes engaging all levels of government, civil society, and the private sector. In keeping with the emerging international consensus on water governance, IWRM, and sustainability principles, South Africa separated the responsibilities for water resources management and water services provision and introduced a new system to allocate water equitably.

A retrospective look at the 14-year strategic planning process that identified the BWP and eventually led to its approval in 2002 reveals that contemporary political, economic, cultural, and technical concerns had to be balanced before a political decision on the dam could be made—especially because this process unfolded during a period of fundamental political transformations.

The decision to proceed with the BWP was reached only after the participatory planning processes, which received considerable media coverage, led politicians to judge that a "sufficient consensus" had been reached. The first enabling step was to move the comprehensive options assessment "upstream" in the planning system. This move was accomplished in public participation processes (PPPs) in the mid-1990s built around the strategic water demand-supply reconciliation analyses of the Berg Water Management Area (WMA). Through the PPPs, representatives of stakeholder interests participated in setting the criteria to select options and the actual evaluation of options.

But it was more than an options assessment. Consistent with the new water laws, the minister for water affairs and forestry advised the National Parliament in 1998 that a positive response to the request by the City of Cape Town (CCT) to proceed with the BWP would be predicated on three actions: (1) a review of Cape Town's water demand projections, (2) a clear indication of the commitment of the CCT and transitional local and district councils to demand-side water management, and (3) better technical information on demand management potential and actual budgetary support for water conservation/water demand management (WC/WDM) programs. This process placed the final BWP decision in a wider sustainability context.

The CCT proceeded to prepare its first water services development plan (WSDP), which incorporated an assessment of demand-side management investments, including the technical, public information and involvement, and tariff measures. In parallel, DWAF moved forward with the project preparation studies. The environmental impact assessment (EIA) featured multistakeholder processes to endorse key project design and operating parameters, with media coverage. In 1999 the Department of the Environment and Tourism (DEAT) issued the Record of Decision (ROD) approving the project EIA. The photos in figure A.1 show the two main physical elements of the BWP during construction in 2006.

The 1999 ROD was unique in the governance-sustainability-communication (G-S-C) linkage in that it set explicit conditions to ensure the ongoing participation of interested and affected parties (I&APs) in the governance arrangements to implement the BWP. A multistakeholder environmental monitoring committee (EMC) was organized around a comprehensive environmental management plan (EMP). Its integral social management aspects ensured that local communities would benefit from project jobs and other employment. The ROD also stipulated that adaptive management of downstream releases was required during the operation phase, involving multistakeholder processes to define and review the monitoring. In fact, the ROD conditions

Figure A.1. Berg River Dam and Supplement Scheme near Franschhoek, South Africa

Berg River Dam near Franschhoek

Berg Supplement Scheme

Photos courtesy of Nigel Rossouw, TCTA Environmental Manager. They appeared in the March 2007 BWP progress report.

encompassed the key issues on which the consensus to develop the project was founded, and the conditions backed those negotiated outcomes with transparent compliance monitoring.

South Africa's cabinet only endorsed the decision to proceed with the BWP in 2002 after the CCT was able to demonstrate that its demand-side management strategy and financing commitments were in place. As a consequence, the BWP emerged as the first bulk water supply project in water-stressed South Africa directly linked to water demand management.

The full case study looks at the critiques of the project and role of the media in public debate around the projects and different perceptions of water security. Environmental and civil rights activist movements abroad and in South Africa, organized under the Skuifraam Action Group (SAG), have pointed to the BWP as a bad decision.[1] Their opposition rested mainly on two grounds: (1) the potential adverse water quality and environment impacts downstream in the Berg River system, and (2) the view that the 1.6 billion rand cost of the BWP would be better invested in WC/WDM.[2]

Viewed through a Governance Lens

The project-level governance arrangements for the BWP were the first to test the implementation of South Africa's new water infrastructure strategy, as well as the working of cooperative governance and partnership approaches at all levels. At the institutional level, the BWP featured the first tripartite public-public partnership to implement a water project in the country, which involved

---

Box A.2.   Linking Options Assessment to Communications, Governance, and Sustainability

In 2003, the World Bank produced a sourcebook on "Stakeholder Involvement in Options Assessment"[a] that was organized around four principles:

- Principle 1: Create an Enabling Environment for Stakeholder Involvement and Options Assessment
- Principle 2: Involve All Relevant Stakeholders
- Principle 3: Assess All Options Strategically and Comprehensively
- Principle 4: Reach a Decision

The sourcebook was aimed at World Bank task teams. It provides guidance on how to improve the involvement of stakeholders in the systematic assessment of options in planning exercises for water and energy services. Ranging from policy formulation to project planning, it draws on ten case studies of programs and projects that considered the development, rehabilitation, and operation of dams around the world. The sourcebook is also intended for water users and stakeholders involved in dam-related planning processes, and for those involved in policy dialogues focused on improvements in participatory planning in the water sector.

Options assessment is clearly important from both sustainability and governance improvement perspectives. Communication capacity is central not only to make the argument for effective involvement of stakeholders in options assessment, but also to put it into practice and thereby promote informed and inclusive decision making.

The Sourcebook on Stakeholder Involvement in Options Assessment and this Communication Handbook are two documents that project managers can bring to the attention of clients considering dam projects in the early stages of project appraisal, especially when the question of good practice is discussed.

a. Energy Sector Management Assistance Program (ESMAP), "Stakeholder Involvement in Options Assessment: Promoting Dialogue in Meeting Water and Energy Needs—A Sourcebook," World Bank, Washington, DC, July 30, 2003.

Cape Town (responsible for water service delivery), the Trans-Caledon Tunnel Authority (TCTA), as project developer and operator, and DWAF as the prime mover (but gradually shifting its role to that of a policy and regulatory body). The project agreements struck were consistent with the philosophy of sustainability in the development and management of dams:

- The Raw Water Supply Agreement (2003) between DWAF and the CCT and the Implementation Agreement (2003) between DWAF and the TCTA reflected the principle that investments in bulk water supply schemes should be recovered from tariffs, and there were sufficient operating budgets for the monitoring and compliance aspects.

- Because of the comprehensive way in which risks were addressed, the international scrutiny of these two agreements by an international credit rating service yielded a highly favorable AA+ credit rating for the BWP debt. The TCTA was then able to secure a mix of concessional and commercial financing from the European Investment Bank (EIB), the Development Bank of South Africa (DBSA), and Absa Bank, a major commercial bank, without a direct government guarantee.

- Critically, the agreements were written to not encumber the adaptive management of the downstream releases from the dam. A key feature of the Water Supply Agreement was the Berg Water Capital Charge (BWCC), which enabled debt to be repaid by the TCTA over 20 years. This fixed charge was added to the existing tariff the CCT paid to DWAF for raw bulk water supplied by the Western Cape system. It was applied from 2004 on—while the BWP was under construction.

For other stakeholders, the EMC brought interested and affected parties into the project governance structure. The committee consisted of an independent chair and elected representatives of the host community and all water user interests in the basin, as well as local government and project authorities. Apart from dealing with environmental concerns, the EMC became the primary platform for dialogue among stakeholders on managing the social components of the project.[3]

Corporate governance was a factor. The TCTA's own corporate governance transition from 2002 was relevant to the BWP governance. It demonstrated the importance of backing progressive legislation, with parallel improvements in the corporate governance of the public enterprises responsible for financing, developing, and operating water infrastructure. For example:

- The TCTA benchmarked its corporate governance policies against national and international good practice. This was the point of entry for improving the transparency, access to information, and anticorruption procedures in the projects it implemented.

- The TCTA developed corporate sustainability reporting and integrated risk assessment procedures as a result of the benchmarking.
  - The TCTA also took specific steps to reduce its corruption risk exposure by adopting the procurement and anticorruption policies required by the funding agencies (EIB and DBSA), which were developing their own corporate policies on fraud and corruption for adoption by all borrowers and contractors.

- Some of the anticorruption steps taken on the BWP reflected the experience of two of the institutional partners (the TCTA and DWAF) in the Lesotho Highlands Water Project, such as requiring contractors to make declarations of any previous fraud or corruption prosecutions.

One indicator of the effectiveness of the combined measures for governance and anticorruption was there was no significant cost increase on the BWP.

An important lesson overall on governance was that integrated risk assessment and management are needed to underpin a shift to integrated water management and the development of sustainable infrastructure. From a communication standpoint, this lesson calls for empowering interested and affected parties in monitoring the risks relevant to their interests and taking steps to bring stakeholder representation into project governance arrangements.

Viewed through a Sustainability Lens

Overall, the BWP reveals how the different aspects of sustainability—environmental, social, governance, economic/financial, physical performance, and institutional—are intertwined. It is not enough to look at each aspect in isolation—for example, environmental sustainability—without considering whether the measures to ensure it are sustainably funded or whether the institutional capacities for monitoring and compliance are present and supported by sufficient resources.

- From a social sustainability perspective, the main concerns on the BWP were addressing divergent perceptions of how the project would affect the host communities in the Franschhoek Valley and La Motte Village, the immediate host. These perceptions were central to gaining local acceptance of the project—on one hand, to minimize the possibility of adverse impacts on heritage, tourism, and the local environment, which are a major part of the local economy, and, on the other, to meet expectations among the previously disadvantaged groups in the valley that they would benefit from the project through jobs and other income-generating opportunities during construction and over the longer term. The project itself had no local water supply dimension. Steps for maximizing local benefits emerged in the multistakeholder discussions during both the project preparation and implementation stages. Among those steps accepted by all parties were the following:
  - *Franschhoek First Policy (FFP).* This policy was aimed at maximizing training and job creation during the construction phase and local procurement of services to ensure that local contractors were able to take maximum advantage of the business opportunities generated by the BWP.

- *La Motte Housing Trust.* Under this mechanism, 80 houses in La Motte Village built under FFP terms for occupancy by the BWP contractor's staff were to be transferred to qualifying local residents after project completion. The proceeds would be used to create a revolving fund for reinvestment in further subsidized housing or in community infrastructure projects to benefit disadvantaged segments of the immediate host community.
- *Sustainable Utilization Plan.* This plan set out longer-term measures for sustainable use of the land surrounding the reservoir and other measures to sustainably integrate the project into the local economy and culture during the operation phase.

Despite the many progressive measures introduced for social sustainability, trust broke down in 2004 between the EMC members on the Franschhoek Valley team, one of the seven subgroups on the EMC, and project authorities. In response, many local community representatives resigned from the EMC, and elected local municipality representatives had to assume these EMC roles directly.

While there were several reasons for the breakdown in trust, it was clear that a segment of the host community felt they would not directly benefit from the project. Moreover, because of past injustices under the apartheid system, it is likely that some members of the host community were seeking some form of reparations, or recognition of their historical marginalization from economic opportunities, through the project.

From the environmental sustainability perspective, the BWP was South Africa's first large dam and water transfer scheme (in the post-1994 era) to provide both low- and high-flow releases to maintain functional aquatic ecosystems downstream within a prescribed river classification system and satisfy the new statutory "reserve" flow for basic human and ecological needs.

The instream flow requirements (IFRs) set for the BWP and the reserve flow determinations for the Berg River (critical stretches) were established in a consultative, transparent process with media coverage.[4] Although there was some criticism of the multistakeholder process and its outcome by SAG members, the IFRs were endorsed by a majority of the downstream water user interest in I&AP workshops. These requirements were designed to achieve an agreed water quantity-quantity "status" downstream and encompassed factors such as salinity management, sediment transport, and river morphology.

The IFRs were key to the final design of the physical structures and operating strategy. For example, they established the size and final design of multilevel intake and associated outlet works and set a limit on the water transfers from the Berg River system to Cape Town water consumers under both normal hydrology and drought conditions. These measures increased the cost of dam construction by about 25 percent.

Although some argue that this cost was unusually high compared with practice in other countries, others argued that the cost was more than justified on the basis of water resource protection considerations and economic values in the basin. The onetime incremental investment of about $5.7 million in the larger intake and outlet structures protected water quality for downstream uses, including irrigators producing high-value export crops worth over $50 million a year—exports mostly destined for European markets, where water quality is critical to meet EU import standards—as well as many direct and indirect jobs.

The BWP thus demonstrates why IFRs should be assessed from a wider water use and economic perspective, engaging with downstream users not just from an ecosystem perspective. Such an assessment is important to making the business case for improvements in sustainability.

Other lessons include the need to ensure that the pre-project baseline river monitoring program gets off to a timely start in order to maximize the years of data gathering before construction and impoundment and to pursue an independent audit of environmental and social management processes. The communication associated with these aspects of the project is important.

Perhaps the most fundamental lesson was the adoption of an adaptive management philosophy for downstream water releases from the BWP linked to monitoring with I&AP input and scrutiny. The ROD stipulated that if monitoring demonstrated the dam was having an unacceptable effect on the river, the release pattern and reserve quantity must be revised.

Viewed through a Communication Lens

South Africa's notion of cooperative governance called for active partnerships between the government and the public to equitably expand access to public services and the provision of infrastructure. Adoption of this more open, inclusive, and, by definition, "communication-intensive" approach in the planning stage was a major factor in gaining public acceptance of the BWP project. It also enriched the design (i.e., improved its environmental and social performance) and laid the foundation for partnership approaches to the implementation and operation phases.

DWAF and the CCT were responsible for public communication during the project identification and preparation stages—that is, up to 2002. The TCTA was responsible for developing the communication strategy for the BWP implementation and operation stages. The overarching communication strategy was structured around three overlapping steams of communication: (1) project communications—disseminating information on the project and implementation issues to the public and stakeholder interests via newsletters, the media, and various information channels; (2) the public participation process (PPP)—facilitating communications integral to I&AP engagement in ongoing decisions about the design and implementation of the project, largely organized around the EMP and social aspects; and (3) EMC communications—facilitating the functioning of the EMC as a stakeholder-driven cooperative governance mechanism and two-way dialogue between EMC members and their constituencies.

The TCTA saw itself as the custodian and facilitator of these three inter-related streams of communication and linked them to its own corporate communication strategy, which provided for reporting on risk management and sustainability issues to BWP stakeholders. What follows are some aspects of the communication strategy:

- To avoid unnecessary confusion and conflict on implementation issues and otherwise provide consistent messages from government, the TCTA developed a mutually agreed communication protocol. One aim was to ensure that the various government actors who communicated with the media and the public (i.e., those from DWAF, the CCT, DEAT, and other agencies) had up-to-date, accurate information.
- Messages were to be clearly set out for each major stakeholder group, and the TCTA was to be responsible for a crisis communication strategy, if required.
- The EMC prepared its own communication protocol that set out how EMC members would interact, including how they would handle matters of internal communication, how they would formalize agreements, and how they would communicate with their respective constituencies and report back to the EMC. The protocol gave the EMC chair sole authority to formally represent the EMC positions to the media, and members themselves were free to present constituency views.

The lessons drawn in the full case study are structured around themes that include the following:

- *Communication strategies must add value.* A communication strategy must fundamentally establish good faith and trust for cooperative governance mechanisms and partnership approaches to work.
- *Interested and affected parties and the public should receive consistent messages and information.* Clear, consistent, reinforcing messages should target the concerns of different I&APs, and they should be issued via the channels preferred by the parties to receive information and have dialogues with partners. This is a policy that all must adopt, including multistakeholder bodies.
- *An overarching sector communication strategy is needed.* The project communication strategy should complement communication strategies that advance the wider reforms.

The BWP story is also interesting to communication practitioners in the sense that it clearly illustrates the practical relevance of the four main branches of modern communication theory and practice in dam projects: development communication, internal communication, corporate communication, and advocacy communication.

Dams as an Integrated Development Intervention

By fortuitous coincidence, Cape Town was host to the World Commission on Dams (WCD) while the BWP was in the project preparation stage. Thus many of the government, civil society, and nongovernmental interests involved in the BWP dialogue also participated in the WCD process from 1998 until the closure of the commission in 2001. Among them was the WCD chair, Kader Asmal, who provided political leadership for South Africa's water governance reforms and the initial decisions on the BWP as minister of water affairs and forestry.

As a result, the BWP was closely compared with the WCD not only by environmental NGOs and civil society organizations (CSOs) opposed to the project, but also by South Africa's government agencies and international funding agencies, including the EIB, which became the single largest project lender. At the end of project implementation, a panel of environmental and social experts will compare the BWP implementation process with the WCD guidelines on dam building.

Together with the South African Multi-Stakeholder Initiative on the WCD, these analyses will provide perhaps the best contextualization of WCD recommendations to a dam project and country policy framework available today.[5] They will reveal the kinds of opportunities and contextual challenges faced in optimizing infrastructure not as a physical asset providing water services, but as a development intervention integrated with the host community's culture and economy.

Conclusion

Although every water project has a unique story to offer, the lessons offered by the BWP are relevant to the challenges that many developing countries are facing today in making choices about water infrastructure.

These lessons reveal that addressing interconnected social and environmental dimensions, together with economic, financial, and technical dimensions, leads to a more sustainable policy framework and infrastructure strategy. They also show how drought and water scarcity affect the public's attitudes toward water storage and dams. On the one hand, the CSOs and environmental NGOs opposed to the BWP felt government communications undermined the effectiveness of environmental opposition to dams. On the other, the government felt the drought simply highlighted public support for the BWP as an integral part of the Western Cape's water security strategy. Proponents of the BWP also argued that the government's delay in the decision to implement the project (from EIA approval in 1999 to cabinet authorization to proceed in 2002) led to demand exceeding supply and the imposition of water restrictions, with subsequent economic losses to various sectors.

But there are in fact two storylines about communications on the BWP. One is about how effective communication underpins public endorsement and the political legitimacy of decision making on bulk water supplies and efforts to make infrastructure more sustainable and developmentally effective. The second is about the role demand management must play in the overall sustainability equation, and the need for investment and communication to work on the behavioral changes necessary. Part of the challenge is to move beyond treating I&AP consultation as a "validation" exercise to actually empowering dialogue mechanisms as learning processes, driving innovative thinking, and pursuing partnership approaches.

The immediate and tangible outcome of the participatory and partnership approaches to the Berg Water Project was that Cape Town benefited from an 18 percent water supply increment in parallel with a 20 percent reduction in projected water demand.

The specific lessons that emerge underline the value that the combination of good governance and genuine I&AP engagement add. The need for effective communication at all levels of decision making and across all stakeholder interests to underpin these accomplishments comes to the forefront.

Notes

1. The Berg River Dam was formerly called the Skuifraam Dam.
2. Those opposed to the Skuifraam Dam argued that a dam option was unnecessary, costly, and environmentally damaging. Instead, a package of water recovery and recycling measures could be mobilized to provide new supplies for immediate and future needs. They argued that the demand management measures introduced were conservative and limited in scope, pointing to achievements of smaller water-stressed municipalities in the region (such as Hermanus) that had mounted programs to reduce peak water demands. They also argued that building the Skuifraam Dam would not change the inequities of water use prevalent in the Western Cape, whereas paying for it would impose high water charges on low-income consumers. Meanwhile, the city charged that the critics were not accurately portraying the situation and that demand and supply interventions were needed.
3. Nevertheless, everything was not perfect; there was a steep learning curve. As noted in the full case study, communication and trust among the Franschhoek Valley members of the EMC broke down, and the local municipalities had to assume a more direct representation role.
4. Some NGOs argue the participation of downstream groups was very weak and that their participation actually undermined the positions of those opposed to the dam.
5. South Africa was a world leader in establishing a multistakeholder national process to assess ways to translate the WCD recommendation into national practices and contexts. See "South African Multi-stakeholder Initiative on the World Commission on Dams," http://www.unep.org/dams/files/SA_Initiative_Final_Report.pdf.

Appendix B
Lesotho Highlands
Water Project

The multipurpose Lesotho Highlands Water Project (LHWP) is designed to transfer water from the water-abundant highlands of Lesotho to the Gauteng region of South Africa (its industrial heartland) and provide hydropower to Lesotho through a series of dams/weirs, delivery tunnels, and associated infrastructure (see box B.1). In addition, for Lesotho one of the primary objectives of the LHWP is to use its export revenues to help alleviate poverty and contribute to economic stability.

To date, Phase I of the LHWP has been completed, as well as the Phase II feasibility study. The responsibilities for these and two more phases are set out in the treaty on the LHWP, which was signed between the Kingdom of Lesotho and the Republic of South Africa in 1986. In the area of environmental and social issues, the treaty requires that (1) all those affected by the project "will be able to maintain a standard of living not inferior to that obtaining at the time of first disturbance," and (2) that implementation, operation, and maintenance of the project should be compatible with protection of the existing quality of the environment and, in particular, should pay due regard to maintaining the welfare of persons and communities affected by the project.

To address widespread perceptions that its institutional arrangements were slow and cumbersome, the governance of the LHWP was set out under Protocol VI to the treaty on the LHWP, which was signed by representatives of both governments in Pretoria on June 4, 1999. Protocol VI provided for

Appendix B originally appeared in Lawrence J.M. Haas, Leonardo Mazzei, and Donal T. O'Leary, *Lesotho Highlands Water Project: Communications Practices for Governance and Sustainability Improvement,* World Bank Working Paper No. 200 (Washington, DC: World Bank, 2010).

> ## Box B.1. Lesotho Highlands Water Project (Phase I): Project Features
>
> Phase I of the project was broken down into two subphases: IA and IB.
>
> Phase IA, which provides for the delivery of 18 cubic meters of water per second, consisted of (1) the 185-meter-high Katse Dam on the Malibamat'so River; (2) 82 kilometers of delivery tunnels to South Africa; (3) the 'Muela Dam on the Liqoe River; and (4) the 72-megawatt 'Muela Hydropower Station. Construction on Phase IA began in 1991, and was commissioned in 1998 at a cost of $2.4 billion.
>
> Phase IB, which provided for the delivery of 11.8 cubic meters of water per second, consisted of (1) the Mohale Dam (9.6 cubic meters per second) on the Senqunyane River; (b) the 15-meter Matsoku Weir (2.2 cubic meters per second) on the Matsoku River and the 6-kilometer delivery tunnel to Katse; and (3) the 32-kilometer delivery tunnel from Mohale to Katse. Final impoundment took place in July 2003 at a cost of $624.3 million.

a structure in which the Lesotho Highlands Water Commission (LHWC) is ultimately responsible for the project but with a shift to a more policy formulation and monitoring role. The board of the Lesotho Highlands Development Authority (LHDA) assumed a greater executive role, but its members would be appointed on the basis of merit by the LHWC, based on a set of proposals of the government of Lesotho (GOL). Protocol VI also provided for the LHDA to assume responsibility for the operations and maintenance of the LHWP within Lesotho and for the Trans-Caledon Tunnel Authority to have similar responsibilities for the project within South Africa. Subsequently, it was agreed that four members of the LHWC would join the LHDA board (which occurred in 2005), although this arrangement was never formalized.

Lessons Learned

Because the LHWP is the largest binational water transfer scheme in the world and because it is divided into phases (as noted, Phase I was divided into two very large subphases, Phase IA and Phase IB, and the feasibility studies for Phase II are under way), the lessons learned in this case study are multifaceted.

Overall perspective.   The formulation and institutional arrangements for the LHWP (particularly for Phase I) have been sufficiently robust to adapt to the major political changes in Lesotho and South Africa.

The project is considered "world class" in terms of the design and implementation of its physical infrastructure, its innovative treatment of environmental flows (EFs), as well as its success in meeting its targets in bulk water supply to South Africa and electricity generation for sale to the Lesotho Electricity Corporation.

The LHWP can serve as a model of mutually beneficial development through demonstrating the benefits of bilateral government cooperation in the development of an international river—benefits that exceed those of individual approaches—as well as in strengthening political cooperation. This model is particularly relevant because about 40 percent of the world's population lives in transboundary river basins, and more than 90 percent of the world's population lives within the countries that share these basins. Within Africa, 81 rivers are shared by two or more countries, well-managed international projects can provide opportunities for alleviating poverty, including through facilitating economic growth.

However, because of its uneven record in addressing its social impacts, partly due to communication defects, the project is still struggling to gain wholehearted support by the host communities and extend benefit-sharing thinking not only between states but among all the stakeholders, including specifically the local communities that host the project and those affected by the resource transformations it causes. It is vital to understand that development of strong political support for these kinds of projects is predicated on their acceptance as development opportunities in which the host communities feel they are full partners, rather than the more traditional acceptance as simply water resources projects developed to meet specific sectoral needs (such as water supply), with the environmental and social impacts appropriately ameliorated.

Governance.   In attempts to address corruption issues, government political will is key. In accordance with the South African Development Community (SADC) Protocol Against Corruption, bribery should be criminalized and vigorously prosecuted. Anecdotal evidence points to the effectiveness of debarment in changing the culture of corruption, particularly in relation to contracts entered into by overseas corporations and developing country agencies, including in the water sector.

However, the focus should be on prevention rather than prosecution. The SADC Protocol Against Corruption sets out a number of preventative measures and mechanisms. According to Transparency International, good operating practice now requires that infrastructure (including water sector) projects include governance improvement plans (GIPs) based on corruption risk assessments at the national, sectoral, and project levels. More support is needed at the project level to develop indicators of corruption. For example, the World Bank has identified the top 10 indicators of project-level fraud and corruption.

Emerging good practice also focuses on the central role that the project developers and proponents can play in combating corruption through the adoption of accepted practices of good institutional governance. A good example is the King Commission (II) report, *Corporate Governance for South Africa* (2002), which articulated a code of good corporate governance that is finding regional acceptance in Botswana (by the Water Utilities Corporation) and South Africa.

Emerging good practice in implementing governance and anticorruption (GAC) strategies on dams is to use a coalition approach in their preparation and implementation. These need to involve all the project stakeholders in different and complementary roles.

Physical sustainability.   To maintain the structural integrity of all its dams, tunnels, and related infrastructure, the LHDA is pursuing a program of activities to be certified under an internationally recognized safety, health, environmental, and quality assurance (SHEQ) risk management program. This plan will enable the LHDA to ensure the optimal transfer/delivery of high-quality water to South Africa and efficient, cost-effective electricity production for Lesotho.

Institutional sustainability. To expedite decision making for such a high-profile project, it would have been appropriate to locate the LHDA within the Prime Minister's Office or under the Council of Ministers rather than having it report to the line ministry. This arrangement could also have enabled the Government of Lesotho to better grasp the development opportunities presented by the project, as well as to improve coordination and management of the transfer of assets once Phase IB of the project was completed.

Ongoing oversight is needed to ensure that the LHDA continues to act transparently and accountably in meeting its responsibilities, particularly in relation to the environmental and social aspects of the project.

Financial sustainability. These lessons learned are related to the water transfer and hydroelectric components of the LHWP.

The financial sustainability of the project's water transfer component is assured by South Africa's continued economic growth and the increasing water demand in the Gauteng region. Revenues are paid from a portion of the Vaal River water user tariff.

Largely because of government inaction on its bulk tariff (which has been pegged at the 2001 level, making it one of the cheapest in the world), the 'Muela hydropower project (MHP) has been lingering in financial uncertainty for the last eight years. This uncertainty has been costly in terms of efficiency, management capacity, and the ability to run the MHP as a commercial entity, including repaying the loans secured to finance its construction.

Environmental sustainability. The Environmental Flows Assessment (EFA) should be conducted in parallel with and as inputs to the Environmental Impact Assessment (EIA) and adequate consultation should be undertaken with other riparians. These activities should happen prior to beginning construction work on dams. Having a policy and legal framework in place to guide the EFA will help to ensure that development-oriented managers will accept the EFA's results.

Sufficient outlet facilities in dams, to accommodate the agreed EFA recommendations, should be incorporated in the design stage and in the project cost estimate (important for financial modeling).

An agreed policy on instream flow requirements (IFRs) to meet a "target ecological condition" of a river immediately downstream of a dam will never fully restore a river to its pristine state. Thus, in accord with good practice, this policy should also include compensation for those affected downstream.

The final agreed IFR scenario (the "Fourth") of LHWP's Phase I was a negotiated outcome that balanced the impacts on downstream users and the losses in royalties and hydropower benefits (which were valued as the wholesale tariff of the then-reliable imports from Eskom, the major South African public electricity utility). But the hydropower benefits of the 'Muela project (due to the inability of Eskom to supply reliable power to Lesotho and the major increases in fuel costs) have radically increased. Thus the negotiated outcome may have to be revisited.

Social sustainability.   In addition to dealing with the project's social impacts on communities downstream from the dam (see the previous section), social sustainability was primarily related to ensuring that the upstream-affected households and communities were treated by the project in accordance with the terms of the LHWP treaty by means of a comprehensive resettlement and compensation program. The major lessons learned were:

- Resettlement housing should be demonstrably superior to the housing lost by those affected, as well as culturally appropriate.
- Compensation programs should consist of a blend of actual compensation (e.g., the project agreed to pay the value of agricultural production over 50 years) and development programs (such as in agriculture, tourism, and small business support).
- It is important to conduct a baseline and regular follow-up surveys and to identify the appropriate key project indicators (KPIs) in order to be able to demonstrate conclusively whether the relevant provisions of the LHWP treaty were met.

Poverty alleviation.   Most of the royalties contributed to economic stability and poverty alleviation because they were direct contributions to government revenues. Nearly two-thirds of the projects funded in the 2004–05 development budget were deemed to be poverty-related.

Two unsuccessful attempts were made to establish a Trust Fund directly linked to the project, the second with the support of the World Bank under a Community Development Support Project (CDSP). The Bank's Project Completion Report for the CDSP was unusually critical of the borrower's and the Bank's performance. Given the core importance of addressing poverty and the critical linkages between large infrastructure development and income restoration/poverty alleviation, successful management of trust funds and similar instruments could be part of a benefit-sharing strategy between the project developers and those affected by the project. Both activities should be closely coordinated throughout the project cycle. Multistakeholder governance for community development funds should be ensured through transparent, accountable processes designed to engage beneficiaries in decisions on the use of funds.

Communication.   Effective communication at all stages of the project cycle (including identification, preparation, implementation, and operation) is critical to the success of complex hydraulic infrastructure projects involving many stakeholders. Communication, properly embedded in the project, is instrumental toward developing consensus on the need for and type of measures to prevent and detect corruption; toward empowering stakeholders in their oversight roles over decision making across the multiple decision points in the life of a project; and toward promoting a culture of disclosure, transparency and accountability.

Key actors often overlooked in any communication strategy are the contractors and other private sector actors, particularly in relation to the interactions of their employees with the host community. As part of the communication strategy, it is critical to identify the possible risk of negative interactions between the contractors' staffs and the local community (e.g., increasing the incidence of communicable diseases such as HIV/AIDS and social tensions or conflict due to language barriers or ethnic, cultural, or religious differences) and put in place a program to minimize the risk.

Effective, responsive complaint management is a critical ingredient of establishing productive relationships between the project developer/sponsor and the host and downstream communities. Although the ombudsman, as an accepted source of appeal, has a critical role to play, the project sponsor continues to retain responsibility for addressing complaints expeditiously—along with, when relevant, contractors and their staffs. Good practice points to providing sufficient resources for this activity, as well as publicly recording complaints and the time frame for their resolution.

Effective communication is a key ingredient of efforts to build support for a sustainable environmental flows policy. Communication is perhaps even more critical in the successful implementation of an EF policy involving an organization's management, dam operators, and those people affected downstream, particularly when high dam flow releases are a factor. Radio has been demonstrated to be an effective communications medium, particularly for isolated, poorer communities.

The LHWP was notable for its progressive learning approach as it moved from the implementation of Phase IA to Phase IB to Phase II. Some examples follow:

- Unlike in Phase IA, in which no formal EIA was undertaken, in Phase IB a complete EIA was conducted, except for an EFA, which was undertaken after the decision to go ahead with the project. For the Phase II feasibility study, a complete EIA (including an EFA) is being undertaken, and the downstream riparians (Botswana and Namibia) are regularly informed of its progress through the Orange-Senqu River Commission (ORSECOM) to which all the riparians belong.

- Budget allocations for the environmental and social components increased from $67 million in Phase IA (about 5 percent of capital costs) to $115.6 million in Phase IB (about 15 percent of capital costs).

- In Phase IA, resettlement of individual households and communities was allowed only within the Katse basin. In Phase IB, to reduce pressure on limited land resources, resettlement was permitted within the Mohale basin and in all Lesotho.

- The steps contributing to better communication in Phase IB of the LHWP than in Phase IA included: (1) appointing community liaison assistants; (2) setting up Community Area Liaison Committees; (3) handling grievances by host and downstream communities, including independent third party adjudication by the Lesotho ombudsman; (4) holding annual stakeholder conferences; (5) opening Public Information Centers at the Katse, Mohale, and 'Muela Dams; (6) setting up a LHWP Web site; and (7) targeting dissemination of World Bank Aide Memoires.
- About 35 percent of the budget for the Phase II feasibility studies was allocated to communication and consultation to engage more fully with local communities and the public and genuinely involve people in decisions that affect them.

World Bank.   The Bank played a vital, long-standing role in facilitating implementation of the LHWP. Its participation began in 1983, when the Bank acted as executing agency for the United Nations Development Programme-financed consultants who supervised the feasibility studies of the LHWP, and continued through various project preparation and supervision stages and then through completion of Phase IB. Although in absolute terms its financial involvement was quite minor (about 5 percent of the Phase I project costs), the Bank financed some of the key strategic components of the project. For example, in Phase IB the Bank financed the engineering design and supervision of the main works; institutional support for the LHDA (including the engineering and environmental and social panels of experts and the Disputes Review Board); and training. The following are some of the lessons learned:

- Through its involvement, the Bank provided comfort to other lenders, including the Development Bank of Southern Africa, the EIB, and export credit agencies, which relied on the Bank's supervision reports to meet their monitoring and evaluation requirements.
- In addressing corruption issues related to Phase IA of the LHWP, the Bank played an important role by debarring two consulting companies that were convicted of bribery in the project: Acres International of Canada for three years as of July 2004 and Lahmeyer International of Germany for seven years as of November 2006. However, the debarment process was slow; 11 months intervened between the conclusion of Acres's appeal of its conviction of bribery in Lesotho and its debarment and 32 months for Lahmeyer. In addition the Bank's own investigation did not find enough evidence to bar the consulting companies; the Bank had to eventually rely on the successful prosecution by the Lesotho authorities for the necessary action.
- The Bank played a successful facilitating role between South Africa and Lesotho in putting in place an IFR policy for the LHWP. It also facilitated an

agreement between Lesotho and South Africa, enabling emergency releases from the LHWP in the event that the flow in the border Caledon River fell to levels that cannot support communities dependent on this source of water supply.

- Through its regular supervision of the project, the Bank's dedicated LHWP Task Team ensured that attention was paid continually to sensitive environmental and social issues through compliance with its safeguard policies.

- However, it seems that political will at the management level was not always as strong as it should have been, particularly in relation to the Community Development Support Project, which had been designed to address the failings of a Trust Fund project and apparently ended up in failure as well. This issue is particularly important in view of the increasing importance of benefit sharing in large dam projects.

Glossary

The following are definitions of some of the terms used frequently in this handbook that may not be familiar to non-Bank readers.

advocacy communication
: Targeted and general communication by any organization to raise awareness of issues and influence relevant policy making or to win public support for a position.

communication-based assessment (CBA)
: The first stage in the four-stage development communication process that selects and applies communication research methods and techniques appropriate to each project and context to understand the sociopolitical context and local dynamics that affect the project objective, design, and stakeholder engagement issues.

communication needs assessment (CNA)
: Within the CBA framework, the CNA is a family of techniques to assess local and institutional communication capacities, understand audiences and media and information environments, and determine information flow patterns and the relevant communication networks.

communication strategy
: A well-planned series of actions aimed at achieving specific objectives through the use of communication methods, techniques, and approaches.

corporate communication
: Communication using media and other methods to portray what an organization does: its code of ethics,

methods of transparency, and trust building with partners. These messages are designed to bridge information and knowledge gaps as well as achieve public confidence.

development communication
A systematic four-phase process of context exploration, context assessment, consensus building among stakeholders, and application of communication media and methods for change to enhance development initiatives and project effectiveness.

internal communication
Communication within an organization or system to keep people informed of issues relevant to the institution and its performance and to ensure the efficient exchange of information among the various units, departments, or staff.

interested and affected parties
"Stakeholders" or "interested and affected parties" refers to individuals and organizations that have an interest in or are affected by an initiative whose outcome they can influence directly or indirectly.

the public
The public is not a homogeneous group of people. Rather, "the public is a constantly shifting multiplicity of affiliations and alliances that group and regroup according to the issues and their understanding of the issues, perceptions of risk and the natural evolution of informal structures" ("Generic Public Consultation Guidelines," issued by the Department of Water Affairs and Forestry, South Africa, 2001).

World Bank project cycle
The framework used by the World Bank to design, prepare, implement, and supervise projects. In practice, the World Bank and the borrowing country work closely throughout the project cycle, although they have different roles and responsibilities.